A guide to the study of

FRESH-WATER BIOLOGY

NEEDHAM and NEEDHAM

A guide to the study of
FRESH-WATER BIOLOGY

A guide to the study of
FRESH-WATER BIOLOGY

Fifth edition, revised and enlarged

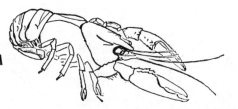

JAMES G. NEEDHAM

PAUL R. NEEDHAM

HOLDEN-DAY, INC., SAN FRANCISCO

Published by Holden-Day, Inc., 500 Sansome Street, San Francisco, Calif. 94111

Fifth Edition

First printing	October, 1962
Second printing	July, 1963
Third printing	June, 1964
Fourth printing	June, 1965
Fifth printing	January, 1966
Sixth printing	September, 1966
Seventh printing	November, 1967
Eighth printing	March, 1969
Ninth printing	July, 1970
Tenth printing	May, 1971
Eleventh printing	April, 1972
Twelfth printing	April, 1973
Thirteenth printing	April, 1974
Fourteenth printing	May, 1975
Fifteenth printing	May, 1976
Sixteenth printing	July, 1977
Seventeenth printing	March, 1978

Library of Congress Catalog Card Number: 62-20742

ISBN: 0-8162-6310-8

Printed in the United States of America

PREFACE

This is the fifth revision of this little book. Its purpose remains the same as that of the earlier editions: to facilitate the recognition of fresh-water organisms both in the field and laboratory. The format remains the same as the earlier editions in offering keys, tables, and figures illustrating mostly genera and in a few cases, species. Only organisms commonly found in fresh waters are included. Inhabitants of salt, brackish, and alkaline waters are omitted. Vascular plants are also omitted. Generic names are applied in an inclusive sense to groups of species that a beginner may be able to recognize by external differences.

The principal change in this edition has been to add the fishes. Keys to the most common forms caught by anglers are presented along with four new plates covering both structures and some 65 different kinds. New plates of protozoans, water bugs, caddis-worm houses, and aquatic beetles are also presented. Numerous revisions have likewise been made in all of the keys.

New materials have been added on the following items: caring for field collections, methods of taking quantitative stream and lake bottom samples, the quantitative distribution and abundance of stream dwelling organisms, methods of making analyses of water for oxygen, carbon dioxide, alkalinities, and the determination of pH values. In addition, references are given to many new books and papers that have appeared since publication of the fourth edition.

For much generous aid in making the revisions, I wish to thank the following colleagues: Dr. William Balamuth (protozoans), Dr. Frank Cole (Diptera), Mr. W. C. Day (mayflies), Dr. D. G. Denning (caddisflies), Mr. W. I. Follett (fishes), Dr. John G. Gallagher (rotifers), Mr. Stanley Jewett (stoneflies), Mr. Hugh B. Leach (aquatic beetles), Dr. Milton C. Miller (crustaceans), Dr. G. F. Pappenfuss (algae), Mr. Allyn G. Smith (molluscs), and Dr. R. L. Usinger (aquatic true-bugs).

For much detailed effort in the arrangement of plates, legends, keys, figure references, and other materials, I am beholden in no small way to Miss Jean Swift of Holden-Day.

I am especially indebted to Mrs. Emily Reid for making the many new drawings required. To Miss Pauline Shorb I owe much for her meticulous care in both typing and proofreading. Graduate student Robert Behnke provided fine assistance in proofreading. Any errors that may be found are solely the fault of the writer.

Observation of aquatic organisms in their natural settings is essential to accompany indoor work on finding out their names. It is simple and easy to tell a mayfly nymph from a stonefly nymph or a water beetle larva from a caddisfly larva. Users of this book should not let the lack of common names, nor the somewhat awe-inspiring scientific names, delay their progress in acquiring a familiarity with one of the richest and most diverse environments in the world. No better teaching materials for young or adults are available or easier to find than aquatic organisms. They are equally useful as research materials and the delights of discovery await all who dip into the water to find them. No niche is too small nor any water too wide or deep, or too hot or too cold, but it will be found to contain its natural faunal elements.

PAUL R. NEEDHAM

Berkeley, California
October 1, 1962

CONTENTS

VIII. STONEFLY NYMPHS (PLECOPTERA)

IX. MAYFLY NYMPHS (EPHEMEROPTERA)

X. DRAGONFLY AND DAMSELFLY NYMPHS (ODONATA)

XI. WATER BUGS (HEMIPTERA)

XII. CADDISFLY LARVAE (TRICHOPTERA)

XIII. TWO-WINGED FLIES (DIPTERA)

XIV. BEETLES (COLEOPTERA)

XV. FRESH-WATER FISHES

Part Two. METHODS OF SAMPLING AND ANALYZING AQUATIC ORGANISMS AND THEIR ENVIRONMENTS

GLOSSARY

GENERAL REFERENCES

Listed below are a number of general references that readers will find helpful in carrying their identification of aquatic plants and animals farther than is possible with this book. In addition, the following references will be found throughout the book on the pages indicated, so that those interested may obtain more detailed and specialized information: Algae, *11;* Protozoans, *14;* Rotifers, *21;* Molluscs and Crustaceans, *29;* Stoneflies and Mayflies, *36;* Dragonflies and Damselflies, *49;* Water Bugs, *52;* Caddisflies, *54;* Two-winged Flies, *55;* Beetles, *65;* Fishes, *78;* and Fishery Biology, *97.*

American Public Health Association
1960   *Standard methods for the examination of water and sewage,* 11th edition. 286 pp. New York, N.Y.

Carpenter, K. E.
1928   *Life in inland waters.* 267 pp. Sidgwick & Jackson, London.

Fassett, N. C.
1940   *A manual of aquatic plants.* 382 pp. McGraw-Hill, New York, N.Y.

Frey, David G. (editor)
1963   *Limnology in North America.* 734 pp. University of Wisconsin Press.

Hyman, Libbie H.
1940–1955 *The invertebrates,* Vols. I-IV. McGraw-Hill, New York, N.Y.

Macan, T. T.
1963   *Freshwater ecology.* 338 pp. John Wiley & Sons, New York, N.Y.

Macan, T. T.
1959   *A guide to invertebrate freshwater animals.* 118 pp. Longmans, Green and Co., London, England.

Macan, T. T., and E. B. Worthington
1951   *Life in lakes and rivers.* 272 pp. London, England.

Morgan, A. H.
1930   *Field book of ponds and streams.* 448 pp. G. P. Putnam's Sons, New York, N.Y.

Muenscher, W. C.
1944   *Aquatic plants of the United States.* 374 pp. Comstock Publishing Co., Ithaca, N.Y.

Needham, J. G., and J. T. Lloyd
1916   *The life of inland waters.* 438 pp. Comstock Publishing Co., Ithaca, N.Y.

Needham, J. G. et al.
1937   *Culture methods for invertebrate animals.* 590 pp. Comstock Publishing Co., Ithaca, N.Y.

Needham, Paul R.
1940   *Trout streams.* 233 pp. Comstock Publishing Co., Ithaca, N.Y.

Pennak, R. W.
1953   *Fresh-water invertebrates of the United States.* 769 pp. Ronald Press, New York, N.Y.

Pratt, H. S.
  1935   *Manual of the common invertebrate animals,* Rev. ed. 854 pp. The Blakiston Co., Philadelphia, Pa.

Ruttner, F.
  1953   *Fundamentals of limnology.* 242 pp. University of Toronto Press.

Smith, G. M.
  1950   *The fresh-water algae of the United States,* Rev. ed. 719 pp. McGraw-Hill, New York, N.Y.

Usinger, R. L., *et al.*
  1956   *Aquatic insects of California with keys to North American genera and California species.* University of California Press, Berkeley, California

Van Reine, W. J. Prud'Homme
  1957   *Plants and animals of pond and stream.* 159 pp. Translated from Dutch by Mona C. Harrison. John Murray, London.

Ward, H. B., and G. C. Whipple
  1959   *Fresh-water biology,* 2nd ed. 1248 pp. Edited by W. T. Edmondson. John Wiley & Sons, New York, N.Y.

Welch, Paul S.
  1952   *Limnology,* Rev. ed. 538 pp. McGraw-Hill, New York, N.Y.

Welch, Paul S.
  1948   *Limnological methods.* 381 pp. The Blakiston Co., Philadelphia, Pa.

Wessenberg-Lund, C.
  1939   *Biologie der Süsswassertiere.* 817 pp. Vienna, Austria.

Whipple, G. C.
  1927   *The microscopy of drinking water.* 323 pp. John Wiley & Sons, New York, N.Y.

A guide to the study of

FRESH-WATER BIOLOGY

Part One

AIDS TO THE RECOGNITION

OF FRESH-WATER ALGAE,

INVERTEBRATES, AND FISHES

P ART 1 consists of various aids to the recognition of fresh-water organisms. The material, arranged in sections in phylogenetic order, includes algae, invertebrates (with particular emphasis on insect larvae) and fishes. Recent studies have added greatly to our knowledge of the immature stages of insects, and the keys in this edition have been correspondingly expanded. In general in headings, common names are given first, followed by the scientific name in parentheses.

Most sections have two aids for recognizing organisms, keys and plates of illustrations. Some have special tables.

KEYS are set up with pairs of opposing characteristics arranged in couplets. To use the key, choose, between the alternatives offered, the one appropriate to the organism being identified; follow the references making similar choices until the name is obtained. In most cases this is the generic name; in some, the family name. Order and class names are in bold face; family names are in small capitals; and names of genera are in italics. The numbers indicating the couplets and the references are italicized; they bear no relation to the figure numbers on the plates. Plate references are given in a separate column following the name. A glossary at the back of the book provides definitions of terms which may be unfamiliar.

PLATES illustrate many but not all of the genera. The names of the genera illustrated are given in alphabetical order in tables at the bottom of the plates. In some of these tables, particularly those for insects and fishes, the following additional information is given. The second column gives the figure number; letters with figure numbers (as 12a) represent details or variants within a genus; numerals within parentheses are used when no whole figure of the genus is given and indicate the figure of another genus similar in form. The third column gives the length of grown specimens, in millimeters (one mm. equals approximately 1/25 inch) or, in the case of protozoans, in microns (μ). The figures for insects are for length of body without antennae and tails. Maturity of nymphs of insects may be judged by length of wing cases. For fishes, the lengths, given in inches, are of mature specimens. The capital letters in the next column indicate continental distribution in a very general way: N, E, S, and W; G, general; C, central. The last column gives habitat in terms of water movement: static (or lentic), and lotic.

I. ALGAE

Neither all algae nor all protozoans may be clearly divided simply into plants and animals. Too much overlap occurs in habits, structures, and their physiology. For this reason, some flagellated forms such as *Chlamydomonas, Volvox,* and *Eudorina* are included here both as algae and protozoans. For a discussion of this problem, see Stanier in Ward and Whipple, *Fresh-water Biology,* 1959, pp. 7-15.

KEY I: ALGAE

Plate
ref.

1 –Cells blue-green; pigments not in chloroplasts.   **IA**
    The blue-greens: Myxophyceae   *2*
   –Cells green, red, or brown   *18*

2 –Cells unicellular or in clusters and colonies, never with cells in filaments; commonly embedded in a gelatinous matrix, more rarely freely floating: **Chroococcales**   *3*
   –Cells (except Spirulina, IA-12) filamentous; branched or unbranched, multiplication by filamentous active hormogones: **Hormogonales**   *9*

3 –Cells solitary or in colonies of less than 50 cells:
    –Cells spherical   *Chroococcus*
    –Cells cylindrical   *Synechococcus, Chroothece*
    –Cells fusiform   *Dactylococcopsis*
   –Numerous cells in gelatinous matrix   *4*

4 –Colonies and gelatine without definite form   *5*
   –Cells in definite arrangement   *6*

5 –Cells in several gelatinous capsules:
    –Cells spherical enclosed in shapeless masses of gelatine   *Gloeocapsa*
    –Cells elongate or elliptical   *Gloeotheca*
   –Cells scattered within the gelatinous matrix:
    –Cells spherical   *Aphanocapsa*   **IA**-13
    –Cells elongate   *Aphanothece*

6 –Colonies free floating   *7*
   –Colonies attached, epiphytes on algae   *Chamaesiphon*

7 –Cell division in 3 planes, colonies therefore in clumps:
    –Cells spherical; colonies spherical when young,
       torn and net-like when older   *Polycystis (Mycrocystis)*   **IA**-8, 9
    –Cells wedge-shaped, colonies spherical   *Gomphosphaeria*
   –Cell division in 2 planes, colonies therefore only one cell deep   *8*

8 –Colonies a hollow sphere without radiating strands   *Coelosphaerium*   **IA**-10
   –Colonies plate-like with rounded cells   *Merismopedia*   **IA**-6

9 –Filaments not attenuated and hair-like at ends   *10*
   –Filaments conspicuously attenuated towards one or both ends: Rivulariaceae. Filaments each with a basal heterocyst; filaments radiating in a gelatinous mass   *Rivularia, Gloeotrichia* (et al)   **IA**-7

10 –Filaments usually not branching   *11*
   –Filaments branching:
    –True branching   Stigonemaceae
    –False branching, heterocysts present   Scytonemataceae

11 –Cells of the filaments all of uniform size, without heterocysts: OSCILLATORI-
    ACEAE   *12*
    –Filaments with occasional cells of different color or larger size (heterocysts):
    NOSTOCACEAE.   *16*

12 –Filaments without sheath   *13*
    –Filaments enclosed in gelatinous sheath   *14*

13 –A single spiral cell   *Spirulina*   **IA-12**
    –Multicellular:
        –Filament a spiral   *Arthrospira*
        –Filament not a spiral   *Oscillatoria* **IA-5**

14 –One filament within a sheath   *15*
    –More than one filament within a thick sheath   *Schizothrix, Microcoleus*

15 –Sheath slimy, filaments often bent, agglutinated   *Phormidium* **IA-1**
    –Sheath firm, not slimy; filaments not in bundles   *Lynbya, Symploca*

16 –Filaments contorted, within a definite gelatinous sheath   *Nostoc* **IA-11**
    –Filament more or less straight, free or in formless slimy mass, without sheath
                                                                            17

17 –Heterocysts terminal; spores long and cylindrical   *Cylindrospermum*
    –Heterocysts not terminal:
        –Filaments aggregated without order   *Anabaena*   **IA-4**
        –Filaments in bundles of plate-like masses   *Aphanizomen* **IA-3**

18 –Organism green, not yellowish green; if reddish then unicellular   *19*
    –Organism yellowish green, red, or brown   *48*

19 –Organism with whorles of leaves: CHARACEAE   *Chara, Nitella*
    –Organism smaller, without whorls: CHLOROPHYCEAE   *20*

PLATE IA: BLUE-GREEN ALGAE

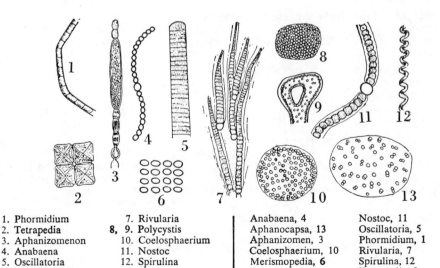

1. Phormidium	7. Rivularia	Anabaena, 4	Nostoc, 11
2. Tetrapedia	8, 9. Polycystis	Aphanocapsa, 13	Oscillatoria, 5
3. Aphanizomenon	10. Coelosphaerium	Aphanizomen, 3	Phormidium, 1
4. Anabaena	11. Nostoc	Coelosphaerium, 10	Rivularia, 7
5. Oscillatoria	12. Spirulina	Merismopedia, 6	Spirulina, 12
6. Merismopedia	13. Aphanocapsa	Polycystis, 8, 9	Tetrapedia, 2

20 –Thallus expanded, membranous   *Ulvales*
    –Thallus neither expanded nor membranous   21

21 –Thallus not filamentous; no conjugation   22
    –Thallus filamentous, though filaments may unite in a plane; if unicellular, conjugation takes place   39

22 –Unicellular or of a definite number of flagellated, motile cells   23   **IIA, IIB**
    –Cells not flagellated or motile (see figs.)   27   **IIA, IIB**

23 –Composed of colonies of many cells with two flagella   24
    –Composed of single cells with 2 or rarely 4 flagella:
        –Contents of cell close to cell wall   *Chlamydomonas*   **IIA-3**
        –Contents of cell connected to cell wall by threads   *Haematococcus*

24 –Colonies spherical or circular   25
    –Colonies flat, cells 4-16 or 32, angles rounded in a colorless sheath   *Gonium*

25 –No gelatinous cover:
        –Many cells in a hollow globe   *Volvox*   **IIA-21**
        –Cells 8-16, arranged in 4 tiers   *Spondylomorum*
    –With a gelatinous cover   26

26 –Colonies round or spherical
        –Cells 16-32-64, globose, not crowded   *Eudorina*   **IIA-7, 8**
        –Cells 4, 8, 16, 32, globose, crowded   *Pandorina*   **IIA-13**
    –Colony of 8 cells in an equatorial zone in a spherical or ellipsoidal covering   *Stephanosphaera*

27 –Cells formed in plates or network: HYDRODICTYACEAE
        –Cells in a flat plate   *Pediastrum*   **IB-1, 2**
        –Cells form a network   *Hydrodictyon*   **IB-25**
    –Cells not in a plate or network   28

28 –Unicellular and solitary; cell with differentiation of base and apex: CHARACIACEAE   *Characium*   **IB-5**
    –Cells without differentiation of base and apex   29

29 –Unicellular and globular or consisting of short, few celled filaments; firm cell wall. Often in damp situations. PLEUROCOCCACEAE   *Protococcus*   **IB-22, 23**
    –Not as above   30

30 –Cells spherical and indefinite in number, embedded in a copious gelatinous envelope   PALMELLACEAE, TETRASPORACEAE   **IB-21**
    –Colonies free or colonial without copious gelatinous envelope; forming autospores   31

31 –Cells elongated, frequently curved; solitary or in definite loosely coherent colonies   32
    –Cells not elongated   34

32 –Colonies enveloped in mucus   *Kirchneriella*   **IB-3**
    –Colonies with little or no mucus   33

33 –Cells attenuated to acute spines:
        –Cells forming definite colonies each of a single row   *Scenedesmus*
        –Cells solitary or loosely grouped in irregular bundles   *Ankistrodesmus*   **IB-11, 12**
        –Cells lunate, arranged back to back   *Selanastrum*   **IB-7**
    –Cells sublunate or ellipsoidal, arranged in groups forming irregular colonies   *Dimorphococcus*

34 –Cells variable, united in a regular flat plate   *Crucigenia*   **IB-26**
    –Cells not united in a flat plate   35

35 –Cells angular with a definite number of angles; cells solitary   *Tetraedron*
    –Cells globose or subglobose   36

PLATE IB: GREEN ALGAE

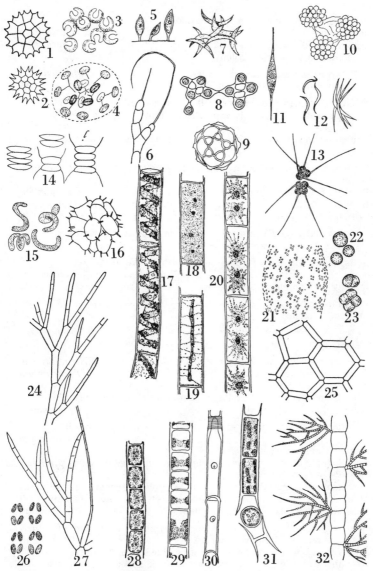

36 –Cells strictly globose, united in a spherical colony:
 –Sphere hollow .... *Coelastrum* **IB**-8, 9
 –Sphere solid, cells with stout spines .... *Sorastrum* **IB**-16
 –Cells not united in a spherical colony *37*

37 –Cells with two or more attenuated bristles .... *Micractinium*
 –Cells without bristles *38*

38 –Cells generally retained within large wall of mother cell OÖCYSTACEAE
 –Cells in grape-like clusters, freely exposed in a thin gelatinous envelope
 .... *Botryococcus* **IB**-10
 –Cells with well-marked subdichotomous connecting threads, chloroplast parietal
 .... *Dictyosphaerium* **IB**-4

39 –Cell division by intercalation of new cells producing transverse striation. **Oedogoniales:**
 –Cells long, filaments unbranched .... *Oedogonium* **IB**-30
 –Cells short with laterally placed bristle, filaments branched *Bulbochaete* **IB**-6
 –Cell division of the ordinary type *40*

40 –Filaments attenuated and commonly ending in a tapered thread *41*
 –Filaments not ending in a tapered thread *46*

41 –Plant of branched filaments forming a flat cushion-like expansion
 .... *Chaetosphaeridium, Coleochaete*
 –Plant entirely filamentous *42*

42 –Filaments branched *43*
 –Filaments not branched *45*

43 –Plants less than 1 cm. high, without setæ .... *Microthamnion*
 –Plants larger, branches attenuated, and with setæ *44*

44 –Filaments fine, showing little difference in character of main axis and branch, not in tufts in gelatinous masses .... *Stigeoclonium*
 –Filaments in tufts in a dense gelatinous mass .... *Chaetophora* **IB**-27
 –Filaments and main branches large, bearing tufts of small branches
 .... *Draparnaldia* **IB**-32

45 –Cells with thick lamellose coats, in a series inside a lamellos sheath
 .... *Cylindrocapsa*
 –Cells without lamellose coat:
  –Chromatophore a homogeneous zonate band, with one to several pyrenoids
 .... *Ulothrix* **IB**-29
  –Chromatophore a parietal disk or plate, with one pyrenoid *Stichococcus*
  –Chromatophore granular, covering more or less completely the whole cell wall, containing starch but no pyrenoids .... *Microspora* **IB**-28

46 –Chloroplast a parietal reticulum with scattered pyrenoids or in form of numerous discs, some containing pyrenoids .... *Cladophora* **IB**-24
 –Chloroplasts single or several, large and of definite shape, with pyrenoids. The entire contents of two cells unite to form a single zygote. **Zygnematales** *47*

47 –Thallus an unbranched thread of many similar cells; sexual reproduction by formation of conjugation tubes between cells of parallel filaments or adjoining cells of the same filament.
 –Chloroplasts of spiral bands .... *Spirogyra* **IB**-17
 –Chloroplasts of two stellate bodies for each cell .... *Zygnema* **IB**-20
 –Chloroplast an axial plate .... *Mougeotia* **IB**-18, 19
 –Unicellular, rarely bound together in a loose thread. DESMIDIACEAE **IC**
 (Desmids, Key IC)

48 –Organism yellowish green. **Xanthophyceæ** *49*
 –Organism grayish or brownish or amber colored *50*

49 –Plants unicellular or forming dendroid colonies   *Ophiocytium*   **IB**-15
    –Plants filamentous, unbranched, cell wall firm, splitting into H-shaped pieces
                                                                                         Tribonema   **IB**-31

50 –Unicellular organisms consisting of 2 silicious valves and two connecting bands
                                                                Bacilliarophyceæ   **ID**
                                                              (Diatoms, Key ID)
    –Cells neither silicious nor 2-valved.   **Chrysophyceæ**
      (A part of this group is included by zoologists in the class **Mastigophora**
      or **Flagellata** among the Protozoa.)

KEY IC: DESMIDS (DESMIDIACEAE)

1 –Cell wall apparently not divided into 2 parts, and without pores   *2*
    –Cell wall showing 2 segments, and with a differentiated outer porous layer   *4*

2 –Cells elongate, cylindrical and not constricted, forming loose filaments. Cell
      wall with a differentiated outer layer, of which the small roughness and
      spines form a part. Chloroplasts axile   *Gonatozygon*   **IC**-1
    –Cells solitary, relatively short and mostly unconstricted   *3*

3 –One chloroplast in each cell:
      –Chloroplast spirally twisted, axile or parietal   *Spirotaenia*   **IC**-2
      –Chloroplast plane, axile, cells solitary   *Mesotaenium*   **IC**-10
    –Two chloroplasts in each cell:
      –Chloroplasts star shaped radiating from a central pyrenoid *Cylindrocystis*   **IC**-25
      –Chloroplasts ridged with longitudinal serrated ridges   *Netrium*   **IC**-12

4 –After division the cell remains free and solitary   *5*
    –After division the cells remain attached to form colonies   *9*

5 –Cells with or without constriction:
      –Cells of moderate length, straight, cylindrical   *Penium*   **IC**-17, 23
      –Cells strongly attenuated towards each extremity, two chloroplasts in each
      cell   *Closterium*   **IC**-5-8
    –Cells more or less constricted at the middle   *6*

6 –Cells elongated and cylindrical, constriction slight:
      –Base of semi-cells plicate   *Docidium*   **IC**-3, 4
      –Base of semi-cells plain   *Pleurotaenium*   **IC**-20
      –Apices of cells cleft, apical incision narrow   *Tetmemorus*   **IC**-19
    –Cells relatively short, deeply constricted   *7*

7 –Cells in vertical view radiating, triangular, quadrangular or radiate, rarely
      fusiform   *Staurastrum*   **IC**-13, 14
    –Cells compressed (at right angles to the plane of the front view), in the
      vertical view fusiform or elliptical   *8*

8 –Cells lobed or incised:
      –Cells mostly oblong or elliptical, moderately lobed   *Euastrum*   **IC**-24, 26
      –Cells very much compressed, deeply lobed or incised   *Micrasterias*   **IC**-15, 18
    –Cells with a more or less entire margin, often furnished with warts or spines
                              Cosmarium   **IC**-21, 22
      –With spines commonly in pairs   *Xanthidium*
      –Spines not in pairs   *Arthrodesmus*

9 –Colonies spheroidal, cells joined by gelatinous bands   *Cosmocladium*
    –Bands very broad, many cells   *Oocardium*
    –Colonies in the form of threads connected by means of apical processes to
      form filaments   *10*

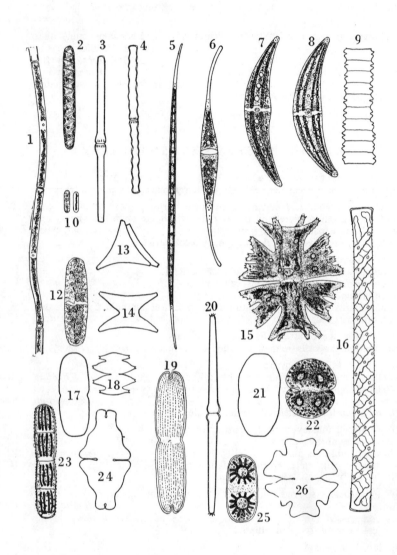

1. Gonatozygon	15. Micrasterias	Closterium, 5, 6, 7, 8	Mesotaenium, 10
2. Spirotaenia	16. Genicularia	Cosmarium, 21, 22	Micrasterias, 15, 18
3, 4. Docidium	17. Penium	Cylindrocystis, 25	Netrium, 12
5, 6, 7, 8. Closterium	18. Micrasterias	Desmidium, 9	Penium, 17, 23
9. Desmidium	19. Tetmemorus	Docidium, 3, 4	Pleurotaenium, 20
10. Mesotaenium	20. Pleurotaenium	Euastrum, 24, 26	Spirotaenia, 2
12. Netrium	21, 22. Cosmarium	Genicularia, 16	Staurastrum, 13, 14
13. Staurastrum (end)	23. Penium	Gonatozygon, 1	Tetmemorus, 19
14. Staurastrum (side)	24. Euastrum		
	25. Cylindrocystis		
	26. Euastrum		

10 –The line of division of the cell where the new and old parts of the cell wall are fitted together, does not develop a girdle during division   *11*
   –The line of division of the cell develops a girdle during division:
      –Cells short, rarely circular with produced angles   *Desmidium*   **IC-9**
      –Cells elongate, cylindrical   *Gymnozyga*

11 –Cells attached by special apical processes:
      –Apical processes very short   *Sphaerozosma*
      –Apical processes long and overlapping the apices of the adjacent cells
                                                                  Onychonema
   –Apices of cells plane and flat:
      –Cells slightly constricted   *Hyalotheca*
      –Cells deeply constricted   *Spondylosium*

KEY ID: DIATOMS (BACILLARIOPHYCEAE)

1 –Cells in transection circular, less commonly polygonal or elliptical, and rarely irregular; valves marked concentrically or radiately by dots, areolations, lines or ribs; cells often with spines, processes or horns. **Centrales 2**
   –Cells in transection narrowly elliptical to linear, less commonly broadly elliptical, lunate, cuneate or irregular valves marked pinnately or transversely by dots, areolations, lines or ribs; cells without spines. **Pennales 5**

2 –Cells short, box-shaped or discoid, mostly circular in transection, usually without horns or projections   *3*
   –Cells of other forms   *4*

3 –Cells forming filaments, girdle sculptured. Valve uniformly marked *Melosira*   **ID-1-4**
   –Cells single, girdle side not sculptured:
      –Without spines   *Cyclotella*   **ID-9, 10**
      –With circle of spines   *Stephanodiscus* **ID-5, 8**

4 –Cells two to many times as long as broad, circular, rarely round, elliptical in transection; girdle with numerous interzones   *Rhizosolenia*
   –Cells box shaped, as long as broad or shorter, elliptical, sometimes lunate in transection; valves with horns, valves with transverse septa, without spines   *Terpsinoe*

5 –Axis of the valves (i.e., the line between the divergent pinnate markings) evident as a narrow strip (pseudoraphe), rarely wanting (i.e., valve without a raphe)   *6*
   –Axis with raphe (slit) present   *11*

6 –Cells usually but little shorter than broad, or longer, with numerous interzones, mostly united into filaments   *7*
   –Cells prevailingly much shorter than broad (rod-shaped, the longer axis of rod representing transverse axis of cell) often united into filaments   *8*

7 –Transverse ribs of valves, when present, not extending into the cell cavity:
      –Valves with a few prominent transverse ribs   *Tetracyclus*
      –Valves finely transversely striate only, pseudoraphe absent *Tabellaria*   **ID-28, 36**
   –Transverse ribs of the valves extending deep into cell   *Denticula*

8 –Cells cuneate in girdle view (i.e., valves not parallel); axis median, interzones present; valves finely transversely striate and with transverse ribs   *Meridion*   **ID-11**
   –Cells rectangular in girdle view, or if cuneate the axis is not median, interzones present or absent   *9*

9 –Axis median without central nodule   *10*
   –Axis near one margin; ends of valves alike:
      –Pseudoraphe and central nodule evident   *Ceratoneis*
      –Pseudoraphe and central nodule not evident   *Eunotia*   **ID-6, 7**

9

PLATE ID: DIATOMS (BACILLARIOPHYCEAE)

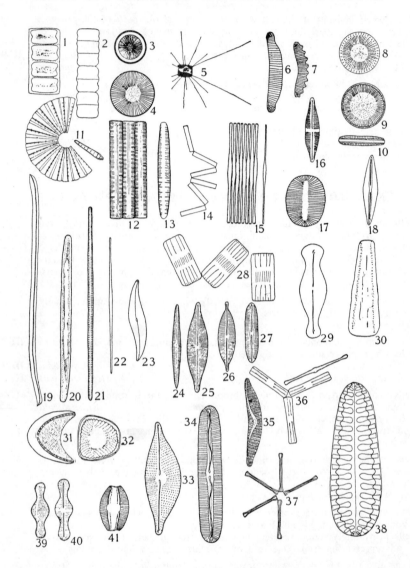

10 –Valves with transverse ribs   *Diatoma*   **ID**-12-14
   –Valves without transverse ribs:
     –Cells arranged radially   *Asterionella*   **ID**-37
     –Cells in filaments or zig-zag chains, valves flat   *Fragilaria*   **ID**-15
     –Cells single or forming fan-like clusters   *Synedra*   **ID**-21, 22

11 –Axis containing an elongate slit (raphe) through cell wall   *12*
   –Axis evident as a narrow, unmarked strip, or keeled; valve with 2 lateral
     wing keels each enclosing a raphe   *21*

12 –Axis commonly median, often more or less lateral, not keeled or when
     keeled not punctate; interzones present or absent   *13*
   –Axis lateral, less often median, punctate; keeled raphe not plainly visible;
     keel at one edge   *Nitzschia*   **ID**-19, 20

13 –Valves alike   *14*
   –Valves not alike   *19*

14 –Valve not keeled   *15*
   –Valve with a keel; keel (including raphe) sigmoid, median   *Amphiprora*

15 –Raphe almost straight   *16*
   –Raphe strongly sigmoid:
     –Cell not twisted   *Gyrosigma*   **ID**-23
     –Cell twisted   *Scoliopleura*

16 –Raphe with simple border   *17*
   –Raphe borderd by two ridges:
     –Central nodule small or slightly elongated   *Brebissonia*
     –Central nodule much elongated, rib-like   *Amphipleura*

17 –Septa of interzones (when present) not fenestrated   *18*
   –Septa of interzones fenestrated:
     –Both valves with a raphe   *Mastogloia*
     –Only one valve with a raphe   *Cocconeis*   **ID**-17

18 –Cells straight in girdle view   *Stauroneis, Frustulia, Pinnularia, Navicula*   **ID**-16, 18
                                                    24-27, 34
   –Cells curved; only one valve with a raphe   *Achnanthes*   **ID**-39, 40

19 –Valves longitudinally symmetrical:
     –Cells straight in girdle view   *Gomphonema*   **ID**-29, 30
     –Cells curved in girdle view   *Rhoicosphenia*
   –Valves longitudinally asymmetrical   *20*

20 –Valves flat, without transverse ribs   *Cymbella*   **ID**-33
   –Valves convex   *Amphora*   **ID**-41
   –Valves with transverse ribs   *Epithemia*   **ID**-35

21 –Valve surface transversely undulate   *Cymatopleura*
   –Valve surface not transversely undulate:
     –Valve cuneate, reniform, elliptical or linear   *Surirella*   **ID**-38
     –Valves subcircular, saddle-shaped   *Campylodiscus*   **ID**-31, 32

REFERENCES: ALGAE

American Public Health Association
   1955   "Biologic examination of water, sewage sludge, or bottom materials." Part 6 of
        Standard methods for the examination of water, sewage and industrial wastes.
        10th ed., 522 pp.

(Algae references continued on page 16)

II. PROTOZOANS

TABLE IIA: PROTOZOANS

Figs.	Genera	Length in μ	Remarks
37.	Acanthocystis	40–100	Usually greenish color, several species
35.	Actinophrys	40–50	Among aquatic plants, often greenish, called "Sun animalcules"
36.	Actinosphaerium	200–300	Among aquatic vegetation
27.	Amoeba	to 600	Fresh or salt water; damp soil
30.	Arcella	30–100	Many species; stagnant water, in bottom ooze and damp soils
1.	Astasia	50–60	Stagnant water; colorless
24.	Bodo	10–15	Stagnant water, often encysted
33.	Centropyxis	100–150	With finger-like cytoplasmic projections; sand grains on body
2.	Ceratium	100–700	Numerous species; fresh and salt water; great color variation
4.	Chilomonas	20–40	Colorless; stagnant water; in bottom ooze and decaying vegetation
3.	Chlamydomonas	10–30	Common in lakes; green
22.	Codosiga	15	Attached by stalks to plants, debris, etc.
31.	Difflugia	200–230	Sand grains encase body; widespread
5.	Dinobryon	30–44	Often colonial, attached to bottom, chromatophores a yellowish brown color
6.	Entosiphon	20	Body oval, flattened, colorless
7.	Eudorina	40–150	Colonies of 16–64 cells; in ponds, lakes and ditches; green
8.	Eudorina	40–150	Colony undergoing asexual reproduction
9.	Euglena	33–55	Abundant in stagnant water with algae. May form green scum on water
32.	Euglypha	20–160	Body covered with siliceous scales
10.	Gonium	90	Forms disc-like colonies, 4–16 cells arranged in single plane; green
26.	Hartmanella	20	A typical soil amoeba
11.	Mallomonas	50	Body covered by siliceous scales
29.	Mayorella	50–300	An amoeba with characteristic tapering pseudopodia
12.	Monas	14–16	Motile in decaying vegetation; colorless
25.	Naegleria	20	Flagellate stage in life cycle of soil-dwelling amoebo-flagellate
23.	Oikomonas	5–20	Stagnant pools and soils; colorless
13.	Pandorina	20–50	In spherical colonies of 8–32 cells in gelatinous mass; ponds and ditches; green
14.	Peranema	40–70	Stagnant water; colorless
15.	Peridinium	44–48	Numerous species; brownish chromatophores
16.	Phacus	40–170	Body flattened and often ridged; many species; very common; green
17.	Pleodorina	450	32, 64 or 128 cells per colony. Somatic and reproductive cells separate; green
18.	Polytoma	15–30	Decaying vegetation in stagnant water; colorless
19.	Synura	100–400	Spherical colonies. Said to impart odor of cucumbers; lakes; golden brown
28.	Thecamoeba	200	An amoeba with a rigid pellicular covering
34.	Trinema	30–100	Among aquatic vegetation
20.	Uroglena	40–400	In ovoid or spherical colonies; gelatinous processes join individual cells
38.	Vampyrella	30–40	Often a bright orange color; predatory on filamentous algae
21.	Volvox	350–500	Colonies of many cells form a hollow ball; green

PLATE IIA: PROTOZOANS

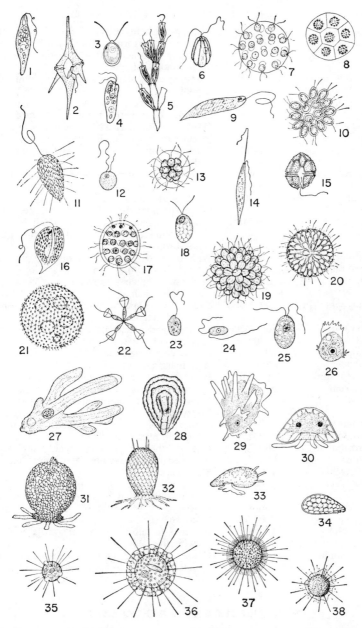

1. Astasia	11. Mallomonas	21. Volvox	30. Arcella
2. Ceratium	12. Monas	22. Codosiga	31. Difflugia
3. Chlamydomonas	13. Pandorina	23. Oikomonas	32. Euglypha
4. Chilomonas	14. Paranema	24. Bodo	33. Centropyxis
5. Dinobryon	15. Peridinium	25. Naegleria	34. Trinema
6. Entosiphon	16. Phacus	26. Hartmanella	35. Actinophrys
7. Eudorina	17. Pleodorina	27. Amoeba	36. Actinosphaerium
8. Eudorina	18. Polytoma	28. Thecamoeba	37. Acanthocystis
9. Euglena	19. Synura	29. Mayorella	38. Vampyrella
10. Gonium	20. Uroglena		

TABLE IIB: PROTOZOANS

Figs.	Genera	Length in μ	Remarks
13.	Blepharisma	80–200	Often pinkish in color; in decaying vegetation
29.	Carchesium	100–125	Many species; forms stalked colonies in which individuals contract separately; some are attached to plants and animals
7.	Chilodonella	50–150	Many species; common surface scum of stagnant pools
21.	Codonella	60–70	Body pot-shaped, sharply divided into collar and bowl; collar without spiral structure.
4.	Coleps	50–110	Many species; characteristic plates covering the body.
9.	Colpoda	40–110	In stagnant pools among decaying vegetation
23.	Cothurnia	70–100	Often in gills of crayfish. Attaches to substrate by short stalk
2.	Dictyostelium	variable	A cellular slime mold; cells of plasmodium distinct, developing into a single sporangium
8.	Didinium	80–200	Predaceous on *Paramecium*
6.	Dileptus	250–500	Many species; with neck-like extension of body
24.	Epistylis	50–250	Colony of many individuals united in a multistalk, not contractile; same species occurs on crayfishes and turtles
18.	Euplotes	90	With isolated groups of compound cilia (cirri)
11.	Frontonia	150–600	Among filamentous algae
20.	Halteria	25–50	Performs bouncing movements. Common in pond-water infusions
31.	Ichthyophthirius	100–1,000	Causes "white-spot" or "Ich" disease of fish in aquaria and fish hatcheries
5.	Lacrymaria	500–1,200	Anterior end extensible and highly flexible
28.	Loxodes	700	Strongly compressed; brownish
16.	Metopus	90–140	In decaying vegetation
17.	Oxytricha	50–250	With marginal rows of cirri
10.	Paramecium	100–350	Many species; very common
1.	Physarum	variable	A true slime mold; amoeba-like cells fuse to form multinucleate plasmodium followed by stalked sporangia
26.	Podophrya	10–100	With sucking tentacles, stalked; close to *Tokophrya*
3.	Prorodon	30–130	A typical, primitive ciliate
14.	Spirostomum	1,000–3,000	One of largest protozoans, highly contractile
15.	Stentor	1,000–2,000	Attached or free-swimming; trumpet-like shape
30.	Stylonychia	100–300	Many species
27.	Tokophrya	50–175	With free-swimming ciliated young; adult stalked, non-ciliated, and bears sucking tentacles
12.	Urocentrum	50–80	Among pond vegetation
19.	Urostyla	200–600	Many species; numerous rows of cirri
22.	Vorticella	135–150	Attaches to substrate by contractile stalk; with free-swimming stage
25.	Zoothamnium	250	Colony of many individuals united in a common stalk; contracts as unit; colonies several mm. high

REFERENCES: PROTOZOANS

Hall, R. P.
  1953   *Protozoology*. Prentice-Hall, Inc., New York, N.Y., 682 pp.

Jahn, Theodore L., and Frances F. Jahn
  1949   *How to know the protozoa*. Wm. C. Brown Co., Dubuque, Iowa, 234 pp.

Kudo, Richard R.
  1954   *Protozoology*. 4th Ed. C. C Thomas, Springfield, Ill., 966 pp.

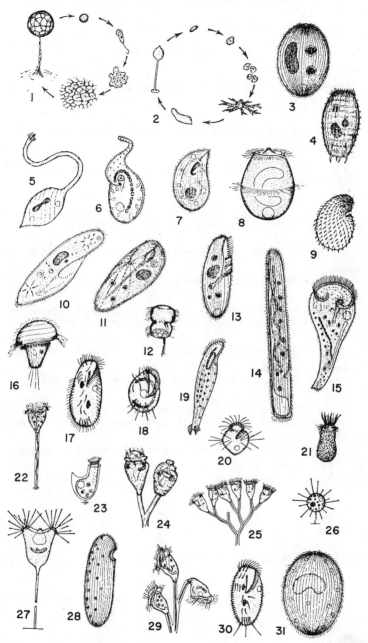

1. Physarum	9. Colpoda	17. Oxytricha	25. Zoothamnium
2. Dictyostelium	10. Paramecium	18. Euplotes	26. Podophrya
3. Prorodon	11. Frontonia	19. Urostyla	27. Tokophrya
4. Coleps	12. Urocentrum	20. Halteria	28. Loxodes
5. Lacrymaria	13. Blepharisma	21. Codonella	29. Carchesium
6. Dileptus	14. Spirostomum	22. Vorticella	30. Stylonychia
7. Chilodonella	15. Stentor	23. Cothurnia	31. Ichthyophthirius
8. Didinium	16. Metopus	24. Epistylis	

REFERENCES: ALGAE

(Continued from page 11)

Cox, C. R.
1952 *Water supply control.* N.Y. State Dept. Health. Bur. Environmental Sanitation. Bull. 22, 279 pp.

Davis, C. C.
1955 *The marine and freshwater plankton.* Mich. State Univ. Press, East Lansing, Mich., 562 pp.

Drouet, F.
1956 *A preliminary study of the algae of northwestern Minnesota.* Proc. Minn. Acad. Sci., 22, pp. 116-138.

Frost, H. S.
1954 *Handbook of algae with special reference to Tennessee and the southeastern United States.* Univ. Tenn. Press, Knoxville, Tenn., 467 pp.

Gainey, P. L., and T. H. Lord
1952 *Microbiology of water and sewage.* Prentice-Hall, Englewood Cliffs, N.J., 430 pp.

Palmer, C. Mervin
1959 *Algae in water supplies.* U.S. Public Health Service Pub. No. 657, 88 pp.

Prescott, G. W.
1951 *Algae of the Western Great Lakes area, exclusive of desmids and diatomes.* Cranbrook Inst. Sci., Bloomfield Hills, Mich., Bull. 31, 946 pp.

Smith, Gilbert M.
1933 *The fresh-water algae of the United States.* McGraw-Hill, N.Y., 716 pp.

Tiffany, L. H., and M. E. Britton
1952 *The algae of Illinois.* Univ. Chicago Press, Chicago, Ill., 407 pp.

Transeau, E. N.
1951 *The Zynemataceae (freshwater conjugate algae) with keys for identification of genera and species.* Ohio State Univ. Press, Columbus, Ohio, 327 pp.

West, G. S., and F. E. Fritsch
1927 *A treatise on the British freshwater algae.* University Press, Cambridge, England, 534 pp.

III. MISCELLANEOUS INVERTEBRATES

1. A hydrachnid, or water mite ×10.
2. A water spider ×1.
3. A gasterotrich, *Chaetonotus* ×10.
4. A coelenterate, *Hydra,* ×10.
5. A tardigrade, *Macrobiotus,* ×20.
6. A bryozoan, *Plumatella* ×3 and ×30.
7. A bristle-worm, *Nais* ×20.
8. A sewage worm, *Tubifex* ×20.
9. A leech, *Placobdella* ×3.
10. A flat-worm, *Planaria* ×5.
11. A colonial rotifer, *Conochilus* ×10.
12. A nematode worm ×10.
13. A fresh-water sponge ×½.
14. Gemules and spicules from the same.

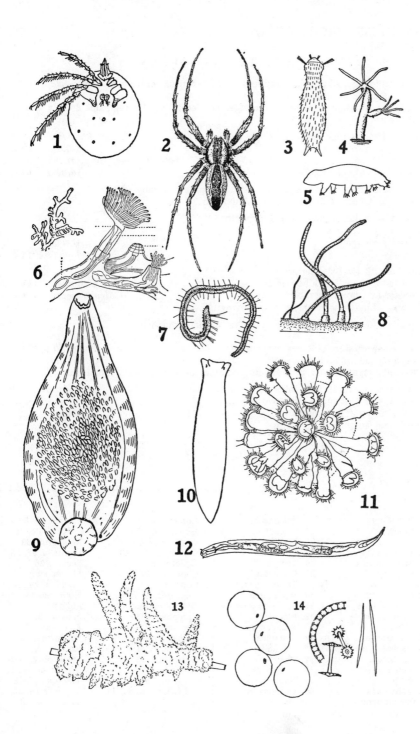

IV. ROTIFERS (ROTATORIA)

KEY IV: ROTIFERS

1 –Head and foot telescopic, rectractile, body ringed, movements leech-like, lateral palpi wanting   **2**
    –Head and foot not telescopic, retractile, not leech-like, lateral palpi usually distinct   **3**

2 –Corona two circular retractile lobes: eye in neck   *Philodina*   **IV-1**
    –Eye in proboscis   *Rotaria*   **IV-2, 3**
    –Eyes absent   *Macrotrachela, Habrotrocha*
    –Corona a flat, ciliated, ventral disc   *Adineta*   **IV-4**

3 –Adult animals attached or in colonies; if separate usually in tubes   **4**
    –Not fixed when adult; rarely in tubular cases   **8**

4 –Corona with long setæ or cilia or both   **5**
    –Corona without setæ:
      –Body elongate   *Acyclus*
      –Body short   *Cupelopagis*   **IV-7**

5 –Corona with long slender setæ:
    –Setæ scattered in lobes of corona   *Collotheca*   **IV-5**
    –Setæ in whorls or rows   *Stephanoceros*   **IV-6**
    –Corona with strong, conspicuous, moving cilia   **6**

6 –Colonial, frequently in spherical clusters: clusters attached
                Lacinularia, Sinantherina
    –Free-swimming colonies or clusters   *Conochilus*   **IV-9**
    –Not attached, inhabiting a tube   *Conochilus*   **IV-9**
    –Attached, separated, or in branching colonies of one to thirty individuals   **7**

7 –Corona of three or four lobes   *Floscularia*   **IV-8**
    –Corona with not more than two lobes:
      –Dorsal antenna present   *Limnias*
      –Dorsal antenna not evident   *Ptygura*

1. Philodina	20, 21. Trichocerca	Adineta, 4	Malleate, 31
2, 3. Rotaria	22. Trichotria	Asplanchna, 13	Forcipate, 39
4. Adineta	23, 24. Mytilina	Brachionus, 30	Kellicottia, 33
5. Collotheca	25. Euchlanis	Chromogaster, 40	Keratella, 35, 36
6. Stephanoceros	26. Lecane	Collotheca, 5	Lecane, 26
7. Cupelopagis	27. Colurella	Colurella, 27	Microcodon, 12
8. Floscularia	28, 29. Testudinella	Conochilus, 9	Mytilina, 23, 24
9. Conochilus	30. Brachionus	Cupelopagis, 7	Notholca, 34
10. Ramate jaws	31. Malleate jaws	Dicranophorus, 19	Philodina, 1
11. Malleo-ramate jaws	32. Platyias	Ephiphanes, 17	Platyias, 32
12. Microcodon	33. Kellicottia	Euchlanis, 25	Ploesoma, 37
13. Asplanchna	34. Notholca	Filina, 16	Polyarthra, 18
14, 15. Synchaeta	35, 36. Keratella	Floscularia, 8	Rotaria, 2, 3
16. Filina	37. Ploesoma	Gastropus, 38	Stephanoceros, 6
17. Epiphanes	38. Gastropus	Hexarthra, 41	Synchaeta, 14, 15
18. Polyarthra	39. Forcipate jaws	Jaws, Ramate, 10	Testudinella, 28, 29
19. Dicranophor-us	40. Chromogaster	Malleo-ramate, 11	Trichocerca, 20, 21
	41. Hexarthra		Trichotria, 22

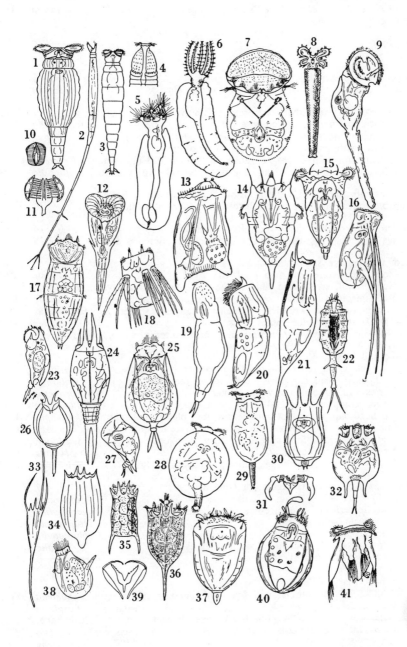

8 –Skipping appendages present: –Three wire-like appendages   *Filina*   **IV-16**
    –Six arm-like appendages   *Hexartha*   **IV-41**
    –Twelve wire-like or blade-like appendages   *Polyartha*   **IV-18**
    –No skipping appendages   9

9 –No foot   10
    –Foot present   12

10 –Stout lorica present   11
    –Cuticle delicate, body large, sac-like, anus absent   *Asplanchna*   **IV-13**

11 –Anus absent: –Lorica oval, compressed dorso-ventrally   *Chromogaster*   **IV-40**
    –Lorica sac-shaped, not compressed   *Ascomorpha*
    –Anus present; no spines, ventral membrane cleft   *Anuraeopsis*
    –Lorica with pattern of longitudinal striations   *Notholca*   **IV-34**
    –Six anterior spines present, lorica dorsum divided into plaques *Keratella*   **IV-35, 3**
    –Three of the six anterior and the single caudal spine very long *Kellicottia*   **IV-33**

12 –Stiff, usually spiny lorica (little distortion on formalin killing)   13
    –Flexible lorica (some or much distortion on formalin killing)   16

13 –Lorica more or less dorso-ventrally compressed   14
    –Lorica spherical, box-like, tubular, or laterally compressed   15

14 –Foot with jointed segments:
    –Body large, one eye, two blade-like toes   *Euchlanis*   **IV-25**
    –Body small, two sharp, dagger-like toes, two lateral eyes   *Lepadella*
    –Body medium to small, one or two toes, one medium eye   *Lecane*   **IV-26**
    –Foot with transverse wrinkles: –Usually six anterior spines   *Brachionus*   **IV-30**
    –Two or ten anterior spines   *Platyias*   **IV-32**
    –No spines, body very compressed dorso-ventrally   *Testidinella*   **IV-28, 2**

15 –Two small toes:
    –With clear cap covering head, lorica bulky, rounded   *Squatinella*
    –With clear head cap, lorica compressed laterally   *Colurella*   **IV-27**
    –Lorica with a row or rows of spines   *Macrochaetus*
    –Lorica compressed laterally, no head cap   *Ploesoma*   **IV-37**
    –No head cap, lorica not compressed, foot placed mid-ventrally *Gastropus*   **IV-38**
    –One or two large toes: –Toes wire-like, sometimes unequal in size *Trichocerca*   **IV-20, 2**
    –Body cylindrical, foot and toes longer than body   *Scaridium*
    –Lorica box-like, ventrally cleft   *Mytilina*   **IV-23, 2**
    –Lorica heavy, form fitting, entirely enclosing body   *Trichotria*   **IV-22**

16 –Body sac-like, bulky, triangular or conical   17
    –Body tubular: –Two long wire-like toes   *Monomatta*
    –Body in series of piled-tire-like folds   *Taphrocampa*
    –With prominent auricles   *Notommata*
    –Dorsal plate split medially   *Cephallodella*
    –One eye, short blade-like toes, auricle-like areas   *Proales*
    –Two eyes, prominent proboscis   *Dicranophorus*   **IV-19**

17 –Body bulky, large or conical:
    –With prominent auricles and four stiff, anterior styles   *Synchaeta*   **IV-14, 1**
    –With auricle-like areas, lorica firm with tail-like projection   *Eosphora*
    –With setae in bunches (3-5) lined across buccal field   *Epiphanes*   **IV-17**
    –With cilia-fringed buccal field and prominent proboscis containing two
       eyes   *Rhinoglena*
    –One eye, conical body tapering regularly to toes, humped   *Cyrtonia*
    –Cuticle delicate, foot very small   *Asplanchnopus*

–Single toe, heart-shaped corona with rotary-like organ attached
                                                 Microcodon   **IV**-12
–Cuticle clear, stomach six-lobed, toes slender   *Enteroplea*
–Corona circular, two spike-like toes set one behind the other dorso-
  ventrally   *Micrododides*

REFERENCES: ROTIFERS

Borntraeger, Gebrüder
  1957   *Rotatoria die Rädertiere Mitteleuropas.* Nickolassee, Berlin, Germany, Vol. 1,
          508 pp., Vol. 2, 115 pp.

Gallagher, John J.
  1957   "Generic classification of the rotifera." *Proc. Penna. Acad. Sci.* Vol. 31, pp.
          182–187.

Harring, Harry K.
  1913   "Synopsis of the Rotatoria." *Bull.* 81, U.S. Nat. Mus., 226 pp.

V. MOLLUSCS (MOLLUSCA)

KEY V: MOLLUSCS

1 –Shell univalve—snails   *2*
    –Shell bivalve—clams   *11*

2 –No operculum   *3*
  –With operculum   *5*

3 –Shell a flat coil, i.e., whorls in same general plane
    –Shell small, up to 6 mm.   *Gyraulus*   **V**-2
    –Shell large, ram's-horn snail   *Helisoma*
    –Shell patelliform, small, depressed   *Ancylus*   **V**-5
    –Shell an elongated spiral   *4*

4 –Shell spire sinistral (to left)   *Physa*   **V**-3
  –Shell spire dextral (to right)   *Lymnaea*   **V**-1

5 –Shell large, spire long, pointed; aperture one-third of length: PLEUOCERIDAE *6*

  –Shell large, not long and pointed; globose, spire short, obtuse, aperture and
    spire about equal in length: VIVIPARIDAE   *7*

  –Shell small, length and width under 11 and 6.5 mm., variable in form   *8*

6 –Shell conic; aperture subrhomboidal, prolonged into a short canal *Pleurocera*   **V**-4
  –Shell slender, ovate-conic, whorls rounded, aperture rounded in front, not
    prolonged into a short canal below   *Goniobasis*   **V**-6

7 –Shell rather thin, globose, whorls convex; animal with foot not produced
    beyond snout   *Viviparus*   **V**-7
  –Shell thick and solid, more elongate whorls slightly convex, foot large,
    much produced beyond snout   *Campeloma*   **V**-8

PLATE V: MOLLUSCS

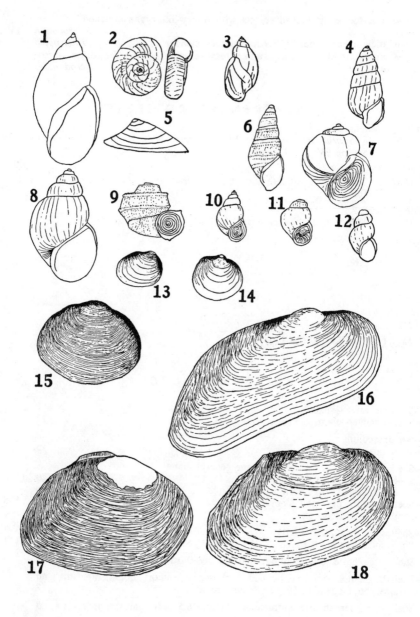

Figs.	Genera	Length	Dist.	Waters		Figs.	Genera	Length	Dist.	Waters
11.	Amnicola	4	G	both		12.	Hydrobia	4	G	both
5.	Ancylus	5	E	both		3.	Physa	16	G	both
18.	Anodonta	90	G	both		13.	Pisidium	8	G	both
10.	Bythinia	10	NE	both		2.	Gyraulus	3–6	G	both
8.	Campeloma	25	N, S, E	both		4.	Pleurocera	30	E, S, W	both
6.	Goniobasis	22	G	both		15.	Sphaerium	14	G	both
1.	Lymnaea	10–50	G	both		17.	Unio	100	E, S, C	both
16.	Margaritifera	30	G	both		9.	Valvata	5	G	both
14.	Musculium	9	G	both		7.	Viviparus	28	E, S	both

8 –Shell flat, discoidal; operculum round, multispiral; aperture round: VALVA-
TIDAE (1 genus)   *Valvata*   **V-9**
 –Shell globose or elongated, length under 10 mm., spire short: AMNICOLIDAE *9*

9 –Aperture oval, length over 7.5 mm.   *Bythinia*   **V-10**
 –Length under 7.5 mm.   *10*

10 –Shell smooth, outer lip of aperture thin   *Amnicola*   **V-11**
 –Shell more slender and elongated; outer lip of the aperture thickened
                                    Hydrobia   **V-12**

11 –Large; length more than 20 mm.   *14*
 –Smaller shells; under 20 mm: SPHAERIIDAE   *12*

12 –Umbones central, shell equilateral   *13*
 –Umbones subterminal, shell inequilateral   *Pisidium*   **V-13**

13 –Shell thin, rounded, smooth, umbones quite prominent   *Musculium*   **V-14**
 –Shell thicker, striate, umbones not so prominent   *Sphaerium*   **V-15**

14 –Shell elongated, laterally compressed, rounded in front, almost lacking a
     posterior ridge. Beak sculpture consisting of a few parallel ridges following
     the growth lines   *Margaritifera**   **V-16**
 –Shell thick, oval to elongate; surface smooth or corrugated   *Unio**   **V-17**
 –Shell thin, beak sculpture consisting of several more or less doubly-looped
     parallel ridges, often slightly nodulous on the loops   *Anodonta**   **V-18**

VI. CRUSTACEANS (CRUSTACEA)

CHART VI A: CLASSIFICATION OF FRESH-WATER CRUSTACEA[†]

Class Crustacea
  Subclass Branchiopoda
    Division Eubranchiopoda
      Order Anostraca (fairy shrimps)
      Order Notostraca (tadpole shrimps)
      Order Conchostraca (clam shrimps)
    Division Oligobranchiopoda
      Order Cladocera (water fleas)
  Subclass Ostracoda
    Order Podocopa (seed shrimps)
  Subclass Copepoda
    Order Eucopepoda (copepods)

  Subclass Branchiura (fish lice)
  Subclass Malacostraca
    Division Peracarida
      Order Mysidacea (opossum shrimps)
      Order Isopoda (aquatic sow bugs)
      Order Amphipoda (scuds, sideswimmers, shrimps)
    Division Eucarida
      Order Decapoda (fresh-water crayfish)

* These were formerly included in one family, the UNIONIDAE, but it has since been split up into many separate families and sub-families. Three common forms are figured here.
† Modified from Pennak's *Fresh-Water Invertebrates of the United States.*

CHART VI B: RECOGNITION CHARACTERISTICS OF FRESH-WATER CRUSTACEA

Malacostraca—body of 20 segments

Names	Plate reference	Form	Eyes	Legs, Pairs	Carapace	Gills	Length
Decapods	VIA—9, 14	Cylindric	Stalked	5	Covers thorax	Under carapace	15-130 mm.
Amphipods	VIA—7, 8	Compressed	Sessile	7	None	Under thorax	5-20
Isopods	VIA—1	Depressed	Sessile	7	None	Under abdomen	5-20
Mysids	VIA—10	Cylindric	Stalked	6	Covers all but 3 segments of thorax	Under carapace	15-30

Branchiopods, Ostracods and Copepods

Names	Plate reference	Form	Eyes	Legs, Pairs	Carapace	Head	Length
Branchiopods	VIA—4, 5, 11	Cylindric	Stalked	19-23	Wanting	Free	25-50
Branchiopods	VIA—2	Depressed	Sessile	21-42	Broad	Consolidated	30-50
Branchiopods	VIA—6	Clam-like	Approximated	13-28	Bivalved	Free, between valves	6-11
Cladocerans (water fleas)	VIB—1-15	Compressed	Sessile	5-6	Bivalved	Free	Microscopic
Copepods	VIB—16, 17, 19, 20	Cylindric	Single	5	Short	Consolidated	Microscopic
Ostracods	VIB—18	Clam-like	Approximated	3	Bivalved	Enclosed	Microscopic

PLATE VI A: CRUSTACEANS: MALACOSTRACANS AND BRANCHIOPODS

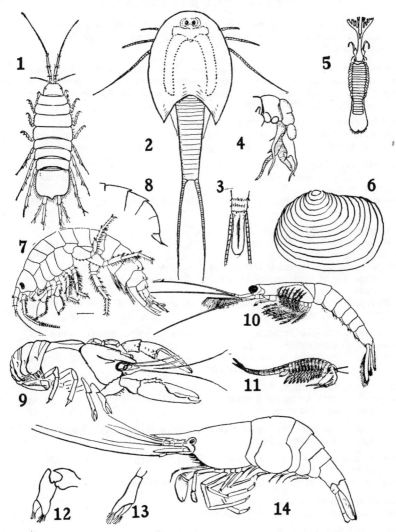

Figs.	Genera	Length	Distr.	Waters		Fig	Genera	Length	Distr.	Waters
(7).*	Anisogammarus	25	W	both		(6).	Lynceus	10	G	static
2.	Apus	30	W	static		10	Mysis	30	G	static
1.	Asellus	15	G	static		(10).*	Neomysis	15	W	static
(9).	Cambarus	80	E	both		(9).	Orconectes	60	G	both
(7).*	Corophium	25	W	both		(9).	Pacifasticus	75	W	both
6.	Estheria	12	W	static		(14).	Palaemonetes	40	S	static
11.	Eubranchipus	25	G	static		(7).	Pontoporeia	11	G	static
(7).	Eucrangonyx	20	G	static		(9).	Procambarus	80	G	both
(1).*	Exosphaeroma	12	W	both		4(11).	Streptocephalus	20	W	static
7.	Gammarus	25	G	static		12 - 14.	Syncaris	30	SW	lotic
8(7).	Hyallella	9	G	static		5.	Thamnocephalus	40	W	static
3(2).	Lepidurus	50	W	static						

* These four forms occur abundantly in coastal stream lagoons on the west coast of the United States. *Corophium* makes small sand tubes in which to dwell; *Anisogammarus* and *Neomysis* are free-swimming forms while *Exosphaeroma* is usually found attached to submerged rocks, logs or other materials.

KEY VI: CRUSTACEANS

The following table includes only the common genera of the free swimming, fresh water crustacea found in the United States. An immature stage, known as **Nauplius** is shown in fig. **VIB**-3.

1 –Without a shell-like covering of the body; with four or five 2-branched swimming feet on the thorax, abdomen without appendages: **Copepoda** *2*   **VIB**-16,
–Usually with a shell-like covering which may or may not entirely cover the   17, 19,
body; if without shell the paired eyes are pedunculate   *5*   20

2 –Cephalothorax distinctly separated from abdomen, the latter small   *3*
–Cephalothorax not sharply differentiated; antennae short, at most 8-segmented (HARPACTICIDAE); with 8-segmented antennae   *Canthocamptus*   **VIB**-19

3 –First antennae long, about as long as the body, 23–25 segmented; in the male, one is modified into a grasping organ; fifth feet not rudimentary: CENTROPAGIDAE   *4*
–Antennæ shorter than cephalothorax; female with 2 egg sacs: CYCLOPIDAE   *Cyclops*   **VIB**-20

4 –Endopodites of swimming feet of 3 segments
   –Antennæ of 23 or 24 segments   *Osphranticum*
   –Antennæ of 25 segments   *Limnocalanus*   **VIB**-17
–Endopodite of first swimming feet of 1 or 2 segments
   –Endopodite of first, second, third and fourth swimming feet of 1 segment   *Epischura*
   –Endopodite of first swimming feet of 2 segments, of third and fourth feet of 3 segments   *Diaptomus*   **VIB**-16

5 –Trunk limbs leaf-like in form: mandible without palp: **Branchiopoda** *6*
–Trunk limbs not leaf-like; mandible with palpus; body not distinctly segmented with caudal furca; antenna large, used for locomotion; bivalve shell, enclosing entire body: **Ostracoda**   *34*   **VIB**-18

6 –With 10 or more pairs of trunk limbs   *7*
–With 4 to 6 trunk limbs; shell bivalved, generally covering body but leaving head free: **Cladocera**   *13*

7 –Without shell; paired eyes pedunculate. The fairy shrimp, etc.: **Anostraca** *8*
–With shell; paired eyes sessile   *10*

8 –Clasping antennæ of male biarticulate. Frontal appendage present: CHIROCEPHALIDAE   *9*
–Clasping antennæ of male triarticulate: STREPTOCEPHALIDAE *Streptocephalus*

9 –Frontal appendage of male rather short   *Eubranchipus*   **VIB**-21
–Frontal appendage of male long and vertical   *Branchinella*

10 –Shell resembling that of a small clam: **Conchostraca**   *11*
–Shell not clam-like   *Notostraca*

11 –Shell spheroidal without lines of growth: LYNCEIDAE   *Lynceus*
–Shell with concentric lines of growth: LIMNADIIDAE   *12*

12 –Shell with 2 to 11 lines of growth   *Eulimnadia*
–Shell with 18 to 22 lines of growth
   –Pediculated dorsal organ on front of head   *Limnadia*
   –No dorsal organ present   CAENESTHERIDAE

13 –Shell restricted to the brood chamber; feet flattened and jointed: *Gymnomera*
   Polyphemus   **VIB**-13
   Leptodora   **VIB**-15
–Shell covering the body; feet not distinctly jointed: **Calyptomera**   *14*

PLATE VI B: CRUSTACEANS: CLADOCERONS, COPEPODS, AND OSTRACODS

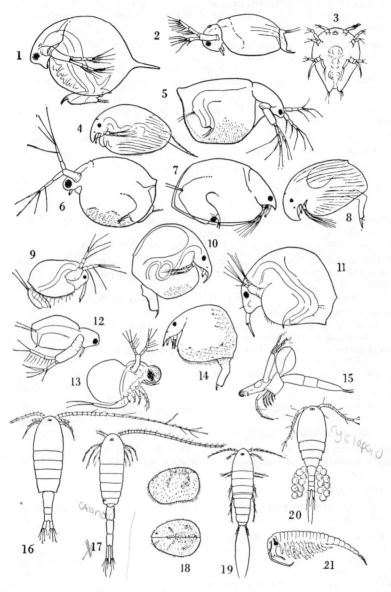

14 –Six pairs of feet; foliaceous: **Ctenopoda**   *15*
–Five or six pairs of trunk limbs, first two pairs more or less prehensile: **Anomopoda**   *19*

15 –Swimming antennæ of female 2-1 branched, the dorsal with many lateral and terminal setæ: SIDIDAE   *16*
–Animal enclosed in a large globular, transparent, gelatinous case open ventrally and forming two valves. Antennæ of female with single ramus: HOLOPEDIDAE   *Holopedium*

16 –Dorsal ramus 3-jointed, ventral ramus 2-jointed; beak present *Sida*   **VIB-2**
–Dorsal ramus 2-jointed; ventral ramus 3-jointed   *17*

17 –Dorsal ramus with interior process or expansion on basal joint *Latona*
–Dorsal ramus without process   *18*

18 –Posterior margin of shell valves with several long setæ:
  –Eyes dorsal far from insertion of antennule   *Latonopsis*
  –Eye ventral or in middle of head   *Pseudosida*
–Posterior margin of shell valves without setæ *Diaphanosoma*   **VIB· 12**

19 –Second antennæ with superior branch 4-segmented; inferior branch 3-segmented   *20*
–Second antennæ with both branches 3-segmented: CHYDORIDAE   *28*

20 –With 5 pairs of feet; first antennæ small, if rarely large (as in Mona) then not inserted at anterior end of head; intestine with 2 hepatic caeca: DAPHNIDAE   *21*
–First antenna long; often with 6 pairs of feet; usually no hepatic caecæ *22*

21 –Head of female with beak:
  –Head keeled above, no transverse suture on neck, shell with polygonal marks and with posterior spine   *Daphnia*   **VIB-1**
  –Spine on shell produced in a straight line with the ventral margin, shell with indistinct net-like markings   *Scapholeberis*
  Spine absent or very short and blunt; markings of transverse lines   *Simocephalus*   **VIB-5**
–Head without a beak:
  –First antenna of female very short, head small and depressed   *Ceriodaphnia*   **VIB-6**
  –First antenna large, head high   *Moina*

22 –First antenna large and fixed; 6 pairs of feet: BOSMINIDAE *Bosmina*   **VIB-11**
–First antenna long and freely movable: MACROTHRICIDAE   *23*

23 –Intestine convolute   *24*
–Intestine simple   *26*

24 –Valves with spines at upper posterior angle   *Ophryoxus*
–Spines absent   *25*

25 –Convolutions of intestine in middle of body:
  –Dorsal margin of shell with conspicuous short backward pointing tooth about the middle   *Drepanothrix*
  –No dorsal tooth   *Parophryoxus*
–Convolutions of intestine in hind part of body:
  –Posterior margin of shell truncated   *Acantholeberis*
  –Posterior margin rounded with slightly pointed protuberance in middle   *Streblocerus*

26 –Upper posterior margin of shell truncated diagonally; general form oval-triangular, the head constituting the apex; ventral and posterior edges of valves enormously dilated   *Ilyocryptus*
–General form ovate, upper posterior margin of shell rounded, not truncated; setæ on ventral margin of shell only   *27*

27 –Setæ long, movable, spine like, and projecting in several directions *Macrothrix* **VIB-9**
–Setæ short and close-set; mid-posterior extremity forming apex of roughly
heart-shaped shell   *Lathonura*

28 –Anus at extremity of post abdomen; 2 hepatic setæ   *Eurycercus* **VIB-7**
–Anus on dorsal side of post abdomen; no hepatic setæ   29

29 –Head and back with high keel; post abdomen very long and slender; with
marginal and lateral teeth   *Camptocercus* **VIB-4**
–Not with all the above characters   30

30 –Hind margin of shell not much less than greatest depth   31
–Hind margin of shell much less than greatest depth   32

31 –Body compressed; claw on concave side with 1 or 2 teeth in the middle:
–Shell with crest   *Kurzia, Acroperus* **VIB-8**
–No crest, post abdomen with marginal and lateral teeth
                                               Alonopsis, Euryalona
–Body not much compressed; claw without tooth or with basal tooth only:
–Beak not produced much beyond first antennæ
                               Oxyurella, Leydigia, Alona, Graptoleberis
–Beak much longer than the first antennæ
                                      Alonella (in part), *Rhynchotalona*

32 –Body noticeably longer than wide *Pleuropus, Alonella* (in part), *Dunhevedia*
–Body spherical or nearly so   33

33 –Post abdomen short with prominent pre-anal angle   *Chydorus*   **VIB-10**
–Post abdomen large, pre-anal angle not conspicuous   *Alonella*   **VIB-14**

34 –Last pair of trunk limbs bent backwards within shell and not used for loco-
motion: CYPRIDIDAE. About 12 Eastern genera
–Broad shell marked dorsally and laterally with 3 prominent dark bands
                                               Cypridopsis vidua   **VIB-18**
–Last pair of trunk limbs directed downwards and used for locomotion:
CYTHERIDAE.
–Free swimming fresh-water form   *Limnocythere*

REFERENCES: CRUSTACEANS AND MOLLUSCS

Baker, F. C.
    1928   *The fresh-water Mollusca of Wisconsin.* Wisc. Acad. Sci. Bull. Part I, Gastro-
            poda, 70:1–505. Part II, Pelecypoda, 70:1–495.

Goodrich, C.
    1932   *The Mollusca of Michigan.* Univ. of Mich. Handbook series, No. 5: 1–120.

Kükenthal, W., and T. Krumbach
    1926–27 *Crustacea.* Handbuch der Zoologie, 3:277–1078.

La Rocque, A.
    1953   *Catalogue of the recent Mollusca of Canada.* Bull. Nat. Mus. Canada, No. 129:
            9–406.

Pelseneer, P.
    1906   *Mollusca.* Part V. of Lankester, *Treatise on zoology,* Adam and Charles Black,
            London, 355 pp.

Schmitt, W. L.
    1938   *Crustaceans.* Smithsonian Scientific Ser., 10:85–248.

Smith, G., and W. F. R. Weldon
    1909   *Crustacea.* Cambridge Nat. Hist., 4:1–217.

Walker, B.
    1918   *A synopsis of the classification of the freshwater Mollusca of North America,*
            north of Mexico. Mus. Zool. Univ. of Mich., Publ: 6:1–213.

INSECTS

KEY TO THE ORDERS OF AQUATIC INSECT LARVAE

1 –Larvæ with wings developing externally (called *nymphs* in this book) and no quiescent pupal stage   *2*
Larvæ proper, with wings developing internally, and invisible till the assumption of a quiescent pupal stage: form more worm-like   *5*

2 –With biting mouth parts   *3*
–Mouth parts combined into a jointed sucking beak, which is directed beneath the head backward between the forelegs: Water bugs   **Hemiptera**   **XI**

3 –With long, slender tails; labium not longer than the head, and not folded on itself like a hinge   *4*
–Tails represented by three broad, leaf-like respiratory plates traversed by tracheæ, or by small spinous appendages; labium when extended much longer than the head; at rest, folded like a hinge, extending between the bases of the forelegs: Dragonflies and damselflies   **Odonata**   **X**

4 –Gills mainly under the thorax; tarsal claws two; tails two: Stonefly nymphs   **Plecoptera**   **VIII**
–Gills mainly on the sides of the abdomen; tarsal claws single; tails generally three: Mayfly nymphs   **Ephemeroptera**   **IX**

5 –With jointed thoracic legs   *6*
–Without jointed thoracic legs; with abdominal prolegs, or entirely legless: Flies, etc.   **Diptera**   **XIII**

6 –With slender, decurved, piercing mouth parts, half as long as the body; small larvæ, living on fresh-water sponges: Sɪsʏʀɪᴅᴀᴇ   **Neuroptera**   **VII**
–With biting mouth parts   *7*

7 –With a pair prolegs on the last segment only (except in *Sialis* **VII-9,** which has a single long median tail-like process at the end of the abdomen) these directed backward, and armed each with one or two strong hooks or claws   *8*
–Prolegs, when present, on more than one abdominal segment; if present on the last segment, then not armed with single or double claws (except in gyrinid bettle larvæ, which have paired lateral abdominal filaments); prolegs often entirely wanting   *9*

8 –Abdominal segments each with a pair of long, lateral filaments: Sɪᴀʟɪᴅɪᴅᴀᴇ   **Megaloptera**   **VII**
–Abdominal segments without long, muscular, lateral filaments, often with minute gill filaments; cylindric larvæ, generally living in portable cases: Caddisfly larvæ   **Trichoptera**   **XII**

9 –With five pairs of prolegs, and with no spiracles at the apex of the abdomen: Moths   **Lepidoptera**   **VII-11**
–Generally without prolegs; never with five pairs of them; usually with terminal spiracles; long, lateral filaments sometimes present on the abdominal segments: Beetles   **Coleptera**   **XIVB**

A spongilla fly, *Sisyra*

RECOGNITION CHARACTERISTICS OF COMMONER FORMS OF AQUATIC INSECT LARVAE

1. Forms in which the immature stages (commonly known as *nymphs*) are not remarkably different from the adults. The wings develop externally and are plainly visible upon the back.

Single distinctive characters are printed in italics.

Common Name and Order	Form	Tails	Gills	Other peculiarities	Habitat	Food habits
Stoneflies (Plecoptera)	depressed	2, long	many, minute, around bases of the legs		rapids	mainly carnivorous
Mayflies (Ephemeroptera)	elongate, variable	3, long: (rarely 2) see gills	7 pairs *dorsal on abdomen*		all waters	mainly herbivorous
Damselflies (Odonata)	slender, tapering rearward		3 leaf-like *caudal gill-plates*	immense grasping lower lip	slow and stagnant	carnivorous
Dragonflies (Odonata)	stout, variable	very short, spine-like	*internal gill chamber* at end of body	immense grasping lower lip	slow and stagnant	carnivorous
Water bugs (Hemiptera)	short, stout, very like adults	variable	wanting	*pointed beak* for puncturing and sucking	all waters	carnivorous

2. Forms in which the immature stages differ very greatly from the adults of the same species, being more or less worm-like, having wings developed internally and not visible from the outside, **and** having the legs shorter, rudimentary, or even wanting (*larvae proper*).

Common Name and Order	Legs	Gills	Rear end of body	Other peculiarities	Habitat	Food habits
Water moths (Lepidoptera)	3 pairs of minute jointed legs followed by a number of pairs of fleshy prolegs	of numerous soft white filaments, or entirely wanting	1 pair of fleshy prolegs with numerous claws on them	claws (crotchets) on all prolegs	all waters	herbivorous
Caddisfly larvae (Trichoptera)	3 pairs rather long	variable or wanting	same as above, with paired larger hooks at tips	mostly living in portable cases	all waters	mostly herbivorous
Orilflies (Neuroptera)	3 pairs shorter	7 pairs of long, lateral filaments tufted at base of lateral filaments, or wanting	*a long tapering tail*		gravelly beds	carnivorous
Hellgrammites, Dobsonflies, Fishflies (Megaloptera)	3 pairs		paired hooked claws		all waters	carnivorous
Water beetles (Coleoptera)	3 pairs	usually wanting	variable	head small often apparently wanting	slow or stagnant	carnivorous
True flies (Diptera)	*wanting*	usually only a bunch of retractile anal gills	see table below		all waters	see table below

3. Further characters of some common dipterous larvae: these are distinguished from aquatic larvae of other groups by the absence of true legs.

Common name and Family	Head	Tail	Fleshy legs, or prolegs	Other peculiarities	Habitat	Food habits
Craneflies (Tipulidae)	retracted and invisible	a respiratory disc bordered with fleshy appendages	variable	*flat lobed body* with *row of ventral suckers*	shoals	mostly herbivorous
Net veined midges (Blepharoceridae)	tapering into body	wanting	wanting	*swollen thoracic segments* "*fans*" on head for food-gathering	rocks in falls	diatoms, etc.
Mosquitoes (Culicidae)	free	with swimming fin of fringed hairs	wanting		pools at surface	herbivorous
Blackflies (Simuliidae)	free	with caudal ventral *attachment disk*	one beneath the mouth		rocks in rapids	herbivorous
True midges (Chironomidae)	free	tufts of hairs	1 *in front* 2 *at rear* end of body	live mostly in soft tubes	all waters	herbivorous
Soldier flies (Stratiomyiidae)	small, free	floating hairs	wanting	depressed form	still water at surface	herbivorous
Horse flies (Tabanidae)	acutely tapering	tapering body	wanting	tubercle covered *spindle shaped body*	beds in pools	carnivorous
Snipe flies (Leptidae)	tapering retractile	with two short tapering tails	stout paired beneath		rapids under stones	carnivorous
Syrphus flies (Syrphidae)	minute	extensile process as long as the body	wanting		shallow pools	
Muscid flies (Muscoidae)	rudimentary	truncated	usually wanting			

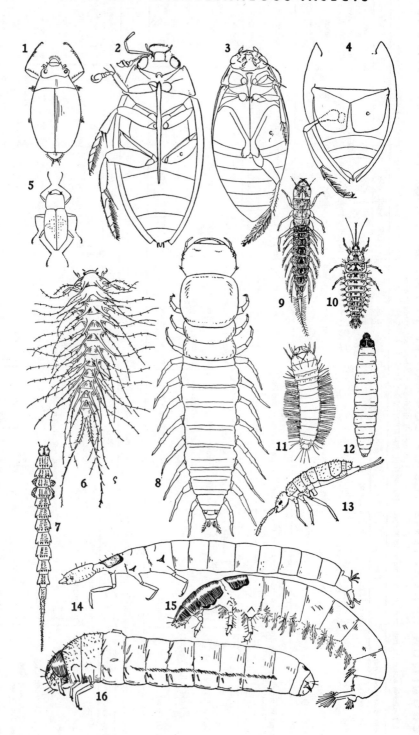

VII. MISCELLANEOUS INSECTS

KEY VII: HELLGRAMMITES
(MEGALOPTERA AND NEUROPTERA LARVAE)

1	–Large forms with biting mouth parts		*2*
	–Small forms with piercing mouth parts: SISYRIDAE		*4*
2	–Body ending in long median tail	*Sialis*	**VII-9**
	–Body ending in a pair of stout hook-bearing prolegs		*3*
3	–Lateral abdominal filaments with a tuft of tracheal gills beneath	*Corydalus*	**VII-(8)**
	–Lateral abdominal filaments with no tracheal gills beneath	*Chauliodes*	**VII-8**
4	–Bristles on back of thorax sessile	*Sisyra*	**VII-10**
	–Bristles on back elevated on tubercles	*Climacia*	**VII-(10)**

VIII. STONEFLY NYMPHS (PLECOPTERA)*

KEY VIII: STONEFLY NYMPHS

1	–With gills on at least some abdominal segments		*2*
	–Without gills (except anal gills) on abdominal segments		*4*
2	–With large single lateral gills on abdominal segments 1 to 7	*Oroperla*	**VIII-(5)**
	–With small tufted ventral gills on the basal segments only		*3*
3	–These gills on segments 1 and 2 only	*Pteronarcys*	**VIII-(5)**
	–These gills on segments 1, 2 and 3	*Pteronarcella*	**VIII-11, (5)**
4	–Body depressed, roach-like in form; head bent under	*Peltoperla*	**VIII-2**
	–Body cylindrical, normal in shape, head directed forward		*5*
5	–First and second tarsal segments very short, together less than half as long as the third segment		*6*
	–First and second segments more than half as long as the third		*19*
6	–With 2 ocelli; copious gill tufts under the thorax		*7*
	–With 3 ocelli; gill tufts variable or wanting		*8*

* See illustration of generalized stonefly nymph on page 36.

Figs.	Adult beetles Family	Figs.	Larvae Genera	Length	Distr.	Waters	Figs.	Genera	Length	Distr	Waters
5.	Dryopidae, Elmidae	8.	Chauliodes	50	G	both	15.	Hydropsyche	18	G	lotic
3.	Dytiscidae	(10).**	Climacia	5	E	lotic	12.§	Nymphula	13	G	static
1.	Gyrinidae	(8).†	Corydalis	75	G	lotic	6.	Peltodytes	10	G	static
4.	Haliplidae	11.‡	Elophila	8	E	lotic	14.	Philopotamus	10	G	lotic
2.	Hydrophilidae	16.	Halesus	18	E	both	9.	Sialis	22	G	static
13.	Collembola (spring-tail)	7.	Haliplus	9	G	static	10.	Sisyra	5	E	lotic

** Differs by having dorsal setae inserted on tubercles.
† Differs by having white gill tufts under lateral abdominal filaments.
‡ Another species of Elophila, lacking tracheal gills, lives in an ovate case on duckweed (*Lemna*).
§ Other species of *Nymphula* have branched tracheal gills.

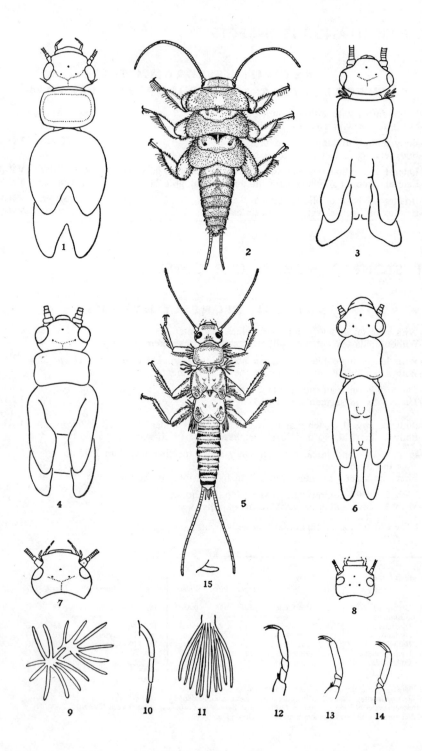

7 –Ocelli close together; anal gills present   *Neoperla*   **VIII-(5)**
  –Ocelli far apart (several diameters of one of them); no anal gills   *Atoperla*   **VIII-(5)**

8 –With conspicuous branched gill tufts at the base of the legs   *9*
  –With gills, if present, minute, unbranched and inconspicuous   *14*

9 –Eyes set far forward, before the middle of the head   *Perlinella*   **VIII-7,**
                                                                      (5)

  –Eyes not before the middle of the head   *10*

10 –Body freckled with small brown dots; length 10 mm   *Perlesta*   **VIII-(5)**
    –Body not so freckled; length at least 20 mm   *11*

11 –With distinct occipital ridge, continuous across the head   *12*
    –Generally with this ridge not continued across the middle   *Acroneuria*   **VIII-(5)**

12 –Anal gills present   *13*
    –Anal gills absent   *Paragentina*   **VIII-(5)**

13 –Abdominal segments yellow, broadly bordered with black *Neophasganophora*   **VIII-(5)**
    –Abdominal segments almost wholly brown   *Claassenia*   **VIII-(5)**

14 –With minute, fingerlike gills under head   *15*
    –Without gills under head   *16*

15 –Gills minute; no gills on thoracic segments   *Isogenus*   **VIII-15,**
                                                       (5)

    –Gills longer; frequently gills also on thoracic segments.   *Arcynopteryx*   **VIII-1,**
                                                       (5)

16 –Head squarish with small eyes set far forward   *17*
    –Head more convex at sides with eyes more prominent laterally   *18*

17 –Pronotal disc heavily fringed with coarse hairs; tails 8 mm. long
                                          Kathroperla   **VIII-(8)**
    –Pronotal disc not heavily fringed with coarse hairs; tails 6 mm. long
                                              Paraperla   **VIII-(8)**

18 –Outer margin of wing pads incurving only at ends; tails longer than either
    abdomen or antennæ   *Isoperla*   **VIII-4**
    –Outer margin of wing pads broadly and regularly rounded; tails shorter than
    either abdomen or antennæ   *Alloperla*   **VIII-(8)**

19 –Second tarsal segment as long as first   *Taeniopteryx*   **VIII-10,**
                                                       12

    –Second tarsal segment shorter than first   *20*

20 –Hind wing pads strongly divergent from body   *Nemoura*   **VIII-3, 9**
    –Hind wing pads nearly parallel with body   *21*

21 –Abdominal segments 1 to 9 divided by a membranous fold laterally *Capnia*   **VIII-(13)**
    –Abdominal segments 1 to 7 divided by a membranous fold laterally *Leuctra*   **VIII-6,**
                                                              14

←―――――――――――――――

Figs.	Genus	Length	Distr.	Waters	Figs.	Genera	Length	Distr.	Waters
(5).	Acroneuria	20	G	lotic	(5).	Neoperla	13	E, S	lotic
(5).	Alloperla	8–12	G	lotic	(5).	Neophasganophora	22	E, S	lotic
1, (5).	Arcynopteryx	20	N, W	lotic	(5).	Oroperla	20	W	lotic
(5).	Atoperla	8	N, E	lotic	(5).	Paragnetina	20	E, S	lotic
(13).	Capnia	6	N, W	lotic	8.	Paraperla	18	W	lotic
(5).	Claassenia	20	N, W	lotic	2.	Peltoperla	6–12	G	lotic
15, (5).	Isogenus	15–20	G	lotic	(5).	Perlesta	12	E, S	lotic
4.	Isoperla	10–15	G	lotic	7, (5).	Perlinella	16	E, S	lotic
(8).	Kathroperla	18	W	lotic	11, (5).	Pteronarcella	18	W	lotic
6, 14.	Leuctra	10	G	lotic	(5).	Pteronarcys	20–30	G	lotic
3, 9.	Nemoura	7–11	G	lotic	10, 12.	Taeniopteryx	12	G	lotic

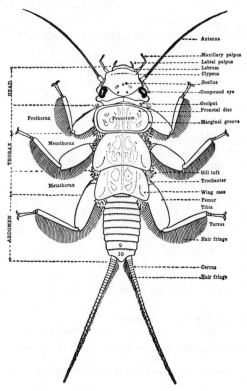

Fig. 1. Generalized stonefly nymph, showing various structures

REFERENCES: STONEFLIES AND MAYFLIES

Claassen, P. W.
  1931   *Plecoptera nymphs of North America.* Thomas Say Foundation of the Entomol. Soc. of Am. Publ. No. 3:1–199, Lafayette, Ind.
  1940   *A catalogue of the Plecoptera of the world.* Cornell Univ. Agri. Exp. Sta. Mem., 232:1-235.

Frison, T. H.
  1935   *Plecoptera of Illinois.* Ill. Nat. Hist. Survey, 20:281-471.

Morgan, A. H.
  1913   A contribution to the biology of mayflies. Ann. Entomol. Soc. Amer., Vol. 6, pp. 371-413.

Needham, J. G.
  1920   Burrowing mayflies of our larger lakes and streams. Bull. U. S. Bur. Fish., Vol. 36, pp. 265-292.

Needham, J. G., and P. W. Claassen
  1925   *The Plecoptera of North America.* Thomas Say Foundation of the Entomol. Soc. of Am. Publ. No. 2:1–397, Lafayette, Ind.

Needham, James G., J. R. Traver and Y. Hsu
  1935   The biology of mayflies. Comstock Publishing Co., 759 pp., Ithaca, N.Y.

Ricker, W. E.
  1952   *Systematic studies in Plecoptera.* Indiana Univ. Publ. Sci. Ser. No. 18:1-200.

IX. MAYFLY NYMPHS (EPHEMEROPTERA)

KEY IX: MAYFLY NYMPHS

1 –Mandibles with a tusk projecting forward and visible from above the head 2
  –Mandibles with no such tusk   9

2 –Tusk flattened, blunt-tipped, and bare (certain Western forms)   **IXB-7,**
        Paraleptophlebia   **(4)**
  –Tusk sharp pointed and more or less hairy   *3*

3 –Fore tibiæ longer than the hind ones   *4*
  –Fore tibiæ shorter than the hind ones   *5*

4 –Mandibular tusks hairy to their tips   *Euthyplocia*   **IXA-1**
  –Mandibular tusks hairy only at their base   *Potamanthus*   **IXA-4**

5 –Front of head with an elevated process   *6*
  –Front of head rounded   *9*

6 –Frontal process rounded   *7*
  –Frontal process bifid at the tip   *8*

7 –Mandibular tusks serrate on the sides   *Ephoron*   **IXA-7**
  –Mandibulbar tusks smooth on the sides   *Hexagenia*   **IXA-6**

8 –Mandibular tusks minutely toothed on the sides   *Pentagenia*   **IXA-3,**
        (6)
  –Mandibular tusks smooth on the sides   *Ephemera*   **IXA-2,**
        (6)

9 –Head strongly depressed: eyes dorsal: gills plate-like   *10*
  –Head not strongly depressed: eyes lateral: gills various   *19*

10 –Gill plates simple, bare   *Arthroplea*
  –Gill plates with clustered gill filaments at the base   *11*

11 –Tails 2   *12*
  –Tails 3   *15*

12 –With a pair of small dorsal spines on each abdominal segment   *Ironodes*   **IXB-(8)**
  –With no such spines   *13*

13 –With a middorsal line of hairs on the abdomen   *Ironopsis*   **IXB-(8)**
  –With no middorsal line of hairs on the abdomen   *14*

14 –Gill plates of first and last pairs directed laterally   *Epeorus*   **IXB-(8)**
  –Gill plates of first and last pairs convergent ventrally   *Iron*   **IXB-(8)**

15 –Gills of segment 7 reduced to tapered filaments   *Stenonema*   **IXB-5**
  –Gills of segment 7 flat like the other gills   *16*

16 –Head emarginate in front: basal gill filaments almost wanting   *Cynigmula*   **IXB-(5,**
        12)
  –Head hardly emarginate: gill filaments well developed   *17*

17 –Gills of segment 1 enlarged, inturned, and meeting beneath the thorax
        Rhithrogena   **IXB-(5)**
  –Gills of segment 1 not enlarged, directed laterally   *18*

18 –Inner canine of mandible about half as long as the outer   *Cynigma*   **IXB-(5,**
        12)
  –Inner canine at least three fourths as long as the outer one   *Heptagenia*   **IXB-5**

19 –Gills completely concealed under an enormously enlarged thoracic shield
        Baetisca   **IXB-10**
  –Gills exposed: thoracic dorsum normal   *20*

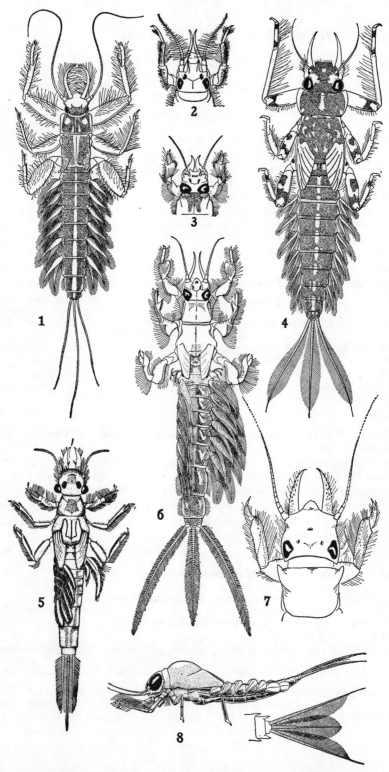

20 –Outer tails fringed alike on both sides	*21*	
–Outer tails heavily fringed on the inner side only	*34*	
21 –Gills present on abdominal segments 1 to 7	*27*	
–Gills wanting from one or more of these segments	*22*	
22 –One pair of gills operculate or gills wanting on abdominal segment 2, gills dorsal	*23* *27*	
–Gills all thin and freely exposed	*Ephemerella*	**IXB-6**
23 –Gill absent on segment 2	*24*	
–Gill present on segment 2	*Oreianthus*	**IXB-(1)**
24 –Hind wing sheath present	*25*	
25 –Gills on segments 2 to 6 double: operculate gill triangular	*Tricorythodes*	**IXB-(1)**
–Gills on segments 2 to 6 single: operculate gill squarish	*26*	
26 –With three prominent tubercles on top of the head	*Brachycerus*	**IXB-(1)**
–With no tubercles on top of the head	*Caenis*	**IXB-1**
27 –Gills of the first pair unlike those that follow	*28*	
–Gills of the first pair similar to the others	*32*	
28 –Gills on segments 2 to 7 single clusters of slender filaments	*Habrophlebia*	**IXB-11, (14)**
–Gills on segments 2 to 7 flattened and more or less plate-like	*29*	
29 –Margins of each gill on middle segments fringed	*Thraulus*	**IXB-(4)**
–Margins of these gills broadly lobed or entire	*30*	
30 –Gill on segment 1 simple; the others lobed at the apex	*Choroterpes*	**IXB-9,(4)**
–Gill on segment 1 double; the others slender tipped	*31*	
31 –Middle gills broad and lobed before the tip	*Blasturus*	**IXB-3,(4)**
–Middle gills narrow and uniformly tapering	*Leptophlebia*	**IXB-7,(4)**
32 –With lateral spines on segments 2 to 9	*Thraulodes*	
–With lateral spines on segments 8 to 9 or on 9 only	*33*	
33 –Hind dorsal margin of segments 1 to 10 finely spinulose	*Paraleptophlebia*	**IXB-7,(4)**
–Hind dorsal margin of segments 7 to 10 only	*Habrophleboides*	**IXB-(4)**
34 –Claws of the fore legs unlike the other claws	*35*	
–Claws of the fore legs like those on the other legs	*37*	
35 –Claws of the fore legs deeply cleft, appearing double	*36*	
–Claws of the fore legs simple, long, slender, bearing a few bristles beneath: front coxæ bearing a thumb-like appendage on the inner side	*Ametropus*	**IXB-(4)**

←

Figs.	Genera	Length	Distr.	Waters	Figs.	Genera	Length	Distr.	Waters
5.	Campsurus	17	S	static	6.	Hexagenia	27	E, C	static*
2, (6).	Ephemera	18	E	both	8.	Isonychia	15	G	lotic
7.	Ephoron	15	E, C	static*	3, (6).	Pentagenia	24	E, C	static
1.	Euthyplocia	29	S		4.	Potamanthus	13	E, C	lotic

*In settling basins in streams conditions are static. Drawings mainly from Kennedy.

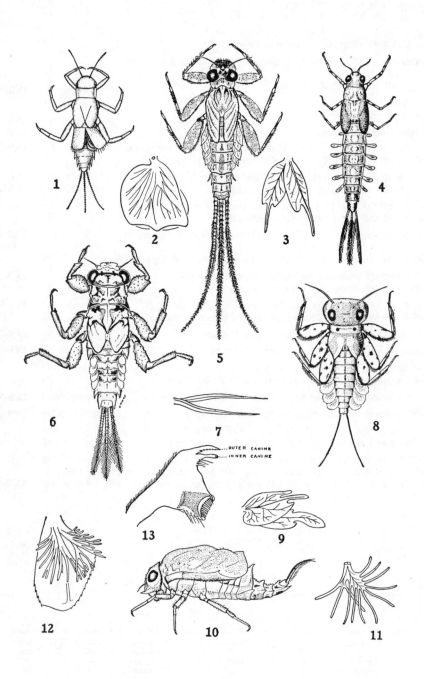

1

2

3

4

5

6

7

8

...OUTER CANINE
...INNER CANINE

13

9

12

10

11

36 –Maxillary palpus 2-jointed; tarsi longer than tibiæ *Metretopus* **IXB-(4)**
 –Maxillary palpus 3-jointed; tarsi at least as long as the tibiæ *Siphloplecton* **IXB-(4)**

37 –Posterolateral angles of the terminal abdominal segments prolonged into thin
 flat lateral spines *38*
 –These angles not greatly prolonged, hardly more than acute *41*

38 –Fore legs conspicuously fringed within by a double row of long hairs: Gills
 on base of maxillæ *Isonychia* **IXA-8**
 –No such hair fringes, and no gills on maxillæ *39*

39 –Gill plates single on abdominal segments 1 to 7 *40*
 –Gill plates double on some of these segments *Siphlonurus* **IXB-2,**
  (4)

40 –Abdominal segments 5 to 9 very wide, onisciform *Siphlonisca* **IXB-(4)**
 –Abdominal segments 5 to 9 normal *42*

41 –Maxillary palpus somewhat pincer-shaped at tip, and end of lacinia beset with
 simple spines *Parameletus* **IXB-(4)**
 –Maxillary palpus normal at tip, and end of lacinia beset with pectinate spines
  Ameletus **IXB-4**

42 –Gill plates folded double on some of the anterior abdominal segments *43*
 –Gill plates all single *45*

43 –Tracheæ of the gill plates pinnately branched: two pairs of wing buds
  Cloeon **IXB-(4)**
 –These tracheæ palmately branched: hind wings buds lacking *44*

44 –Gill flap folded over ventral surface: maxillary palpus 2-jointed *Callibaetis* **IXB-(4)**
 –Gill flap folded over dorsal surface of the gill plate: maxillary palpus 3-
 jointed *Centroptilum* **IXB-(4)**

45 –Tails 2 *46*
 –Tails 3 (except in the Western *Baetis bicaudatus*) *47*

46 –Hind wing buds present (very minute) *Heterocloeon* **IXB-(4)**
 –Hind wing buds lacking *Pseudocloeon* **IXB-(4)**

47 –Middle tail shorter than the others (except in *B. tricaudatus*) *Baetis* **IXB-(4)**
 –Middle tail as long as the others: last joint of labial palpus widely dilated *48*

48 –Hind wing buds present *Centroptilum* **IXB-(4)**
 –Hind wings buds absent: gill tracheæ branched to one side *Neocloeon*

Figs.	Genera	Length	Distr.	Waters		Figs.	Genera	Length	Distr.	Waters
4.	Ameletus	9	G	both		(4).	Heterocloeon	4	N, E	static
(4).	Ametropus	15–17	SW	lotic		(8).	Iron	9	G	lotic
(4).	Baetis	6	G	lotic		(8).	Ironodes	15	W	lotic
10.	Baetisca	10	E, C	lotic		(8).	Ironopsis	15	NW	lotic
(1).	Brachycercus	5–7	G	static		7, (4).	Leptophlebia	7	G	both
3, (4).	Blasturus	9	E	both		(4).	Metretopus	10	N	static
1.	Caenis	3	G	static		(1).	Oreianthus	14–17	SE	lotic
(4).	Callibaetis	9	G	static						
(4).	Centroptilum	4	G	static		7, (4).	Paraleptophlebia	5–9	G	both
(4).	Cloeon	8	G	static		(4).	Parameletus	11–12	N	static
9, (4).	Choroterpes	7	G	lotic		(4).	Pseudocloeon	2–4	G	static
(5, 12).	Cynigma	8–9	NW	lotic		(5).	Rhithrogena	10	W	lotic
(5, 12).	Cynigmula	7	N, W	lotic		(4).	Siphlonisca	19–20	NE	static
8.	Epeorus	8	G	lotic		(4).	Siphloplecton	11–16	N, E	static
6.	Ephemerella	8–15	G	both		2, (4).	Siphlonurus	11	G	static
11, (4).	Habrophlebia	5	E	lotic		5.	Stenonema	6–14	G	lotic
(4).	Habrophlebiodes	6	E	static		(4).	Thraulus	6	N, W, S	static
5.	Heptagenia	10	G	lotic		(1).	Tricorythodes	4	G	lotic

X. DRAGONFLY AND DAMSELFLY NYMPHS (ODONATA)

Fig. 2. Diagrams explaining terms used in the dragonfly key: *A* is a labium (of *Perithemis*); *ml* is its median lobe; *ll*, its lateral lobe; *ms*, mental setae; *ls*, lateral setae; *h*, hook.

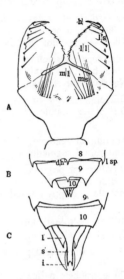

    B (*Perithemis*) and *C* (*Anax*) are end of abdomen as viewed from above; 8, 9, and 10 are the three last segments; *dh*, dorsal hooks; *l sp*, lateral spines; *l*, lateral, *s*, superior, and *i*, inferior appendages.

KEY X: DRAGONFLY AND DAMSELFLY NYMPHS

1 –Gills within the stout spine-tipped abdomen (suborder **Anisoptera:** dragonflies)   2
  –Gills, three flat vertical plates at the end of the slender abdomen (suborder **Zygoptera:** damselflies)   43

2 –Labium flat   3
  –Labium spoon-shaped, covering the face up to the eyes   20

3 –Antennæ short and thick   4
  –Antennæ slender and bristle-like: AESCHNINAE   13

4 –Antennæ seven-jointed; PETALURINAE   *Tachopteryx* **XA-1**
  –Antennæ four-jointed: third joint longest   5

5 –Tenth abdominal segment as long as all other segments combined   *Aphylla* **XA-2**
  –Tenth abdominal segment no longer than other single segments   6

6 –Middle legs closer together at base than are the fore legs   *Progomphus* **XA-8**
  –Middle legs not closer together at base than are the fore legs   7

7 –Wing cases strongly divergent   *Ophiogomphus, Erpetogomphus* **XA-4, (4)**
  –Wing cases parallel   8

8 –Abdomen and third antennal segment flat and nearly circular   *Hagenius* **XA-9**
  –Abdomen and third antennal segment more elongate   9

9 –Third antennal segment thin, flat, oval   10
  –Third antennal segment more nearly cylindric   11

10 –Third antennal segment broadly oval   *Lanthus* **XA-11**
    –Third antennal segment elongate   *Octogomphus* **XA-10, (11)**

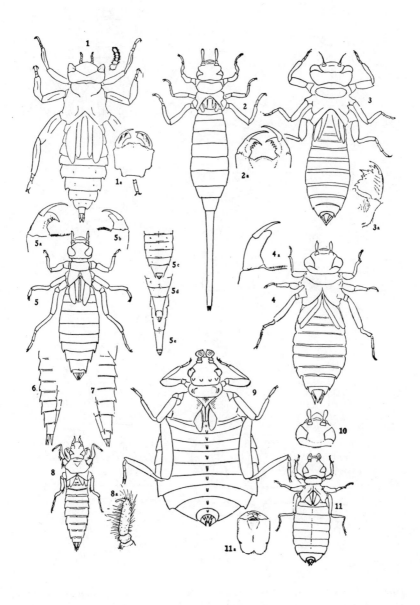

Figs.	Genera	Length	Distr.	Waters	Figs.	Genera	Length	Distr.	Waters
2.	Aphylla	45	S	static	(4).	Erpetogomphus	27	W	lotic
3.	Cordulegaster	42	G	lotic	11.	Lanthus	22	E	lotic
7, (5).	Dromogomphus	33	E, C	lotic	4.	Ophiogomphus	27	G	lotic
5.	Gomphus	24–45	G	both	10, (11).	Octogomphus	23	W	lotic
6, (5).	Gomphoides	34	SW	?	8.	Progomphus	30	G	lotic
9.	Hagenius	40	E	static	1.	Tachopteryx	38	E	static

PLATE XB: DRAGONFLY NYMPHS

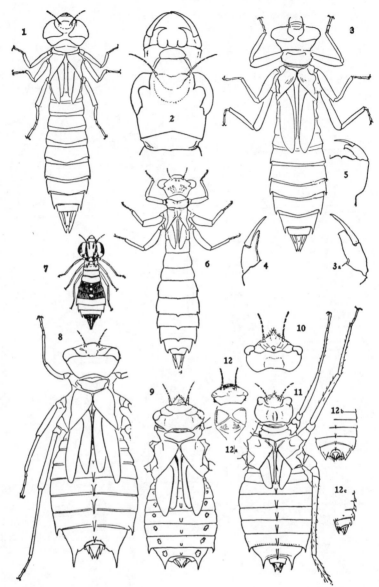

Figs.	Genera	Length	Distr.	Waters	Figs.	Genera	Length	Distr.	Waters
1.	Aeschna	45	G	static	6.	Nasiaeschna	38	E, C	both
7,* (1).	Anax	45	G	static	11.	Didymops	28	E	static
4, (3).	Basiaeschna	42	E	lotic	8.	Epicordulia	27	E, S	static
3.	Boyeria	38	E	lotic	12.	Helocorulia	20	E	static
2.	Coryphaeschna	52	SE	static	10, (11).	Macromia	32	G	static
5.	Epiaeschna	48	E, C	static	9.	Neurocorduleia	21	E, C	both
(1).	Gynacantha	45	S	static	(8).	Tetragoneuria	22	G	static

* A young nymph; the bands of color later disappear.
The first 8 are Aeschninæ, with flat labium. The remainder are Cordulinæ with spoon-shaped labium.
The lateral spines of the abdomen are in No. 2 on segments 7–9; in 1, 5 and 7, on 6–9; in 4. 6 and 8, on 5–9; and in No. 3, on 4–9.

11 –Dorsal hooks on abdominal segments 6–9 long and sharp   *12*
   –Dorsal hooks on abdominal segments 6–9 short and blunt   *Gomphus*   **XA-5**

12 –Lateral abdominal appendages as long as inferiors   *Gomphoides*   **XA-6, (5)**
   –Lateral abdominal appendages shorter than inferiors   *Dramogomphus*   **XA-7, (5)**

13 –Lateral lobes of labium armed with strong raptorial setæ   *Gynacantha*   **XB-(1)**
   –Lateral lobes of labium lacking raptorial setæ   *14*

14 –Hind angle of head strongly angulate   *15*
   –Hind angle of head broadly rounded   *17*

15 –Superior abdominal appendages as long as inferiors   *Coryphaeschna*   **XB-2**
   –Superior abdominal appendages much shorter than inferiors   *16*

16 –Lateral lobe of labium squarely truncate on tip   *Boyeria*   **XB-3**
   –Lateral lobe of labium with taper pointed tip   *Basiaeschna*   **XB-4, (3)**

17 –Lateral spines on abdominal segments 7–9   *Anax*   **XB-7, (1)**
   –Lateral spines on abdominal segments 6–9   *Aeschna*   **XB-1**
   –Lateral spines on abdominal segments 4 or 5–9   *18*

18 –Low dorsal hooks on segments 7–9   *Nasiaeschna*   **XB-6**
   –No dorsal hooks on abdomen   *Epiaeschna*   **XB-5**

20 –Inner edge of lateral lobe of labium coarsely and irregularly toothed
             Cordulegaster   **XA-3, 3a**
   –Evenly and regularly toothed or entire   *21*

21 –Head with a high frontal horn   *22*
   –Head smooth or with a low rounded prominence   *24*

22 –Dorsal abdominal hooks sharp, flat and cultriform   *23*
   –Dorsal abdominal hooks thick and blunt   *Neurocordulia*   **XB-9**

23 –Head widest across the eyes   *Macromia*   **XB-10, (11)**

   –Head widest across the bulging hind angles   *Didymops*   **XB-11**

24 –Dorsal hook on the ninth abdominal segment   *25*
   –No dorsal hook on the ninth abdominal segment   *29*

25 –Lateral setæ of labium 5   *26*
   –Lateral setæ of labium 7   *27*

26 –Lateral spines of 9 surpassing tips of terminal appendages   *Epicordulia*
   –Same shorter not surpassing tips of terminal appendages   *Perithemis*   **XC-15**

27 –Lateral spines of 9 surpassing tips of terminal appendages   *Tetragoneuria*   **XB-(8)**
   –Same, shorter not surpassing tips of terminal appendages   *28*

28 –Length under 15 mm   *Helocordulia*   **XB-12**
   –Length over 20 mm   *Somatochlora*   **XC-7, (6)**

29 –Eyes at sides of head   *30*
   –Eyes capping anterolateral angles of head; more frontal than lateral   *40*

30 –Abdominal appendages strongly decurved: lateral spines wanting   *Erythemis*   **XC-11**
   –Abdominal appendages straight or but a very little declined; lateral spines on
       8 and 9   *31*

31 –Dorsal hooks present   *32*
   –Dorsal hooks absent   *34*

32 –Spines of segment 9 very long   *Celithemis*   **XC-10, (9)**

   –Spines of segment 9 short or moderate   *33*

45

33 –Dorsal hooks as long as the segments which bear them		
	Leucorrhrinia, Dythemis	**XC-9, (9)**
–Dorsal hooks shorter than the segments which bear them	*Sympetrum**	**XC-14**
34 –Abdomen smooth		*35*
–Abdomen hairy		*38*
35 –Lateral spines very short	*Paltothemis*	**XC-(12)**
–Lateral spines long		*36*
36 –Spines of 8, short; of 9, long	*Pachydiplax*	**XC-12**
–Spines of 8 and 9 both long and similar		*37*
37 –Teeth on inner edge of lateral lobe of labium deeply cut	*Pantala*	**XC-5, (4)**
–Teeth obsolete	*Tramea*	**XC-4, 4a**
38 –Lateral setæ 6	*Nannothemis*	**XC-13**
–Lateral setæ 7		*39*
–Lateral setæ 8–10	*Erythrodiplax*	**XC-(14)**
–Lateral setæ more than 10	*Sympetrum**	**XC-14**
39 –Lateral spines of 9 one fifth as long as 9	*Cordulia*	**XC-6**
–Lateral spines of 9 one third as long as 9	*Dorocordulia*	**XC-(6)**
40 –Lateral setæ 0–3	*Ladona*	**XC-8, (1)**
–Lateral setæ 5–10		*41*
41 –Median lobe of labium evenly contoured	*Libellula*	**XC-1**
–Median lobe of labium crenulate on front border		*42*
42 –Lateral setæ 8	*Orthemis*	**XC-2, (1)**
–Lateral setæ 10	*Plathemis*	**XC-3, (1)**

DAMSELFLY NYMPHS (ZYGOPTERA)

43 –Basal segment of antennæ as long as all others together		*44*
–Basal segment of antennæ not longer than other single segments		*45*
44 –Median lobe of labium cleft below bases of lateral lobes	*Agrion*	**XD-1**
–Median lobe of labium cleft only to base of lateral lobes	*Hetaerina*	**XD-11(1)**
45 –Labium spoon-shaped, narrowed in the middle		*46*
–Labium regularly tapered backward to middle hinge		*47*
46 –Mental setæ of labium 5 each side	*Lestes*	**XD-3**
–Mental setæ of labium 7 each side	*Archilestes*	**XD-4 (3)**
47 –Gills half as broad as long: no mental setæ		*48*
–Gills not more than one third as broad as long: mental setæ present		*49*
48 –Lateral setæ 1–4	*Argia*	**XD-2**
–Lateral setæ, none (occasionally 1)	*Hyponeura*	**XD-(2)**
49 –Hind angles of head angulate		*50*
–Hind angle of head rounded		*51*
50 –Gills widest in middle, one third as broad as long	*Amphiagrion*	**XD-7**
–Gills widest toward the distal end, one sixth as broad as long	*Chromagrion*	**XD-5, (7)**
51 –Mental setæ of labium 1 or 2	*Nehalennia*	**XD-9, (6)**
–Mental setæ 3 to 7		*52*

* A new genus, *Tarnetrum,* has been erected for two species of *Sympetrum, T. corruptum* and *T. illotum,* both widely ranging, whose nymphs entirely lack dorsal hooks on the abdomen. In nymphs of the former species the mental setae number 17 and the lateral setae, 13-14. In *T. illotum* the mental setae are 13 and lateral setae, 9.

PLATE XC: DRAGONFLY NYMPHS

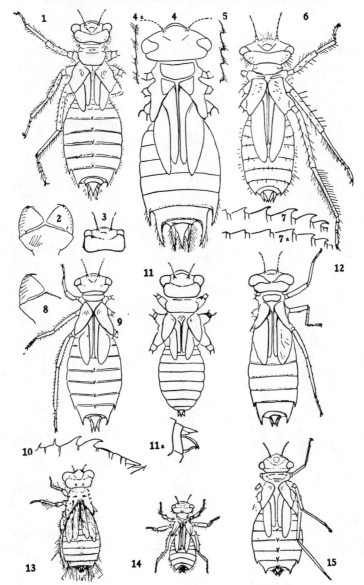

Figs.	Genera	Length	Distr.	Waters	L. Setæ*
10, (9).	Celithemis	15–21	N, E, S	static	8–9
6.	Cordulia	22	N	static	7
(6).	Dorocordulia	20	N	static	7
(9).	Dythemis	17	SW	lotic	10
11.	Erythemis	17	G	static	8
(14).	Erythrodiplax	12–14	E, S	static	8–10
8, (1).	Ladona	22	E	static	6
9.	Leucorrhinia	18	E, N	static	10–11
1.	Libellula	22–27	G	static	5–10

Figs.	Genera	Length	Distr.	Waters	L. Setæ
13.	Nannothemis	10	E, N	static	6
2, (1).	Orthemis	22	SW	static	8
12.	Pachydiplax	18	G	static	10
(12).	Paltothemis	23	SW		9
†5, (4).	Pantala	25	G	static	12–14
15.	Perithemis	15	SE	static	15
3, (1).	Plathemis	24	G	static	10
7, (6).	Somatochlora	26	E, N	static	6–7
14.	Sympetrum	13	G	static	9–14
4, 4a†.	Tramea	25	G	static	10–11

* *L. Setæ* = the raptorial setæ on the lateral lobe of the labium, within.
† This is the toothed inner border of the lateral lobe of the labium: compare fig. 4a of plate.

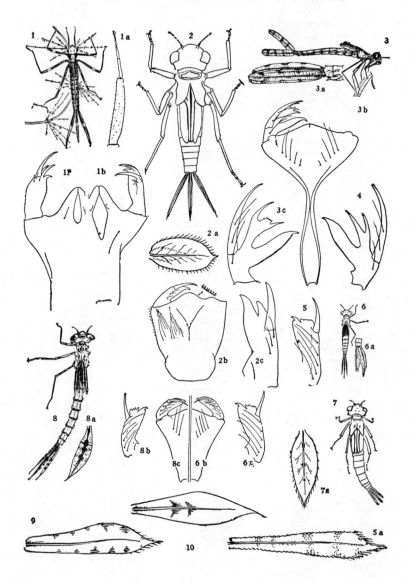

Figs.	Genera	Length	Distr.	Water	Setæ* L	Setæ* M
1.	Agrion	20	G	lotic	0	0
7.	Amphiagrion	11	G	static	5–6	4
10, (6).	Anomalagrion	8	E, S	static	5	4
4, (3).	Archilestes	30	C, W	static	3	7
2.	Argia	12–17	G	both	1–4	0
(8).	Coenagrion	13	N	static	6–7	4–6
5, (7).	Chromagrion	12	E	static	5	3–4
8.	Enallagma	12–15	G	both	4–5	2–4
(8).	Hesperagrion	15	SW	static	6	3–4

Figs.	Genera	Length	Distr.	Waters	Setæ* L	Setæ* M
11, (1).	Hetaerina	18	E, S, W	lotic	0	0
(2).	Hyponeura	22	SW	lotic	0–1	0
6.	Ischnura	11–13	G	static	5–6	4
3.	Lestes	20	G	static	3	5
9, (6).	Nehalennia	11	G	static	6	1
(6).	Teleallagma	18	SE	static	6	3
(6).	Telebasis	14	SW	static	6–7	2
(8).	Zonagrion	14	SW	static	5	3

* Raptorial setæ inside the labium; L, lateral; M, mental setæ.

XI. WATER BUGS (HEMIPTERA)

52 –Gills ending in slender tail-like tips	*53*	
–Gill tips blunt or merely acute	*55*	
53 –Gills with about 6 crossbars of brown	*Zoniagrion*	**XD-8**
–Gills with fewer crossbars or with none	*54*	
54 –Length of body 18 mm	*Teleallagma*	**XD-(6)**
–Length of body 11 or 12	*Ischnura*	**XD-6**
55 –Notch in upper margin of middle gill beyond its middle	*56*	
–This notch before middle of gill or wanting	*57*	
56 –Width of gill more than a third of its length	*Telebasis*	**XD-(6)**
–Width of gill less than a third of its length	*Coenagrion*	**XD-(8)**
57 –Gill tips rounded	*Hesperagrion*	**XD-(8)**
–Gill tips angulate or slightly pointed	*58*	
58 –Gills not pigmented; slender	*Anamalagroin*	**XD-10, (6)**
–Gills more or less strongly pigmented	*Enallagma*	**XD-8**

REFERENCES: DRAGONFLIES AND DAMSELFLIES

Nymphs have been described for a number of genera that are not included in this key. Useful references covering these and other forms are as follows:

Gloyd, Leonora K., and Mike Wright
 1959 "Odonata." In Ward and Whipple, *Fresh-water biology*, 2nd ed., W. T. Edmond-son, ed. pp. 917-940 John Wiley & Sons, Inc., New York.

Needham, J. G., and Minter Westfall
 1955 *Dragonflies of North America*. Univ. of Calif. Press, Berkeley, Calif., 615 pp.

Walker, E. M.
 1953 The *Odonata of Canada and Alaska*. Vol. I, Part I, General Part II, The Zygoptera-damselflies, Univ. of Toronto Press, 292 pp.

XI. WATER BUGS (HEMIPTERA)

KEY XI: WATER BUGS (HEMIPTERA)*

1 –Antennæ shorter than head	*2*
–Antennæ as long as or longer than head, exposed	*6*
2 –Hind tarsi with indistinct setiform claws (except in Plea which is less than 3 mm. long)	*3*
–Hind tarsi with distinct claws	*4*
3 –Head overlapping thorax dorsally. Front tarsi 1-segmented, palæoform	CORIXIDAE
–Head inserted in thorax. Front tarsi normal: NOTONECTIDAE	*10*
4 –Membrane of hemelytra (front wings) reticulately veined	*5*
–Membrane of hemelytra without veins: NAUCORIDAE	*12*

* After Hungerford, Kansas Univ. Sci. Bull., Vol. 11, 1919.

PLATE XI: WATER BUGS (HEMIPTERA)

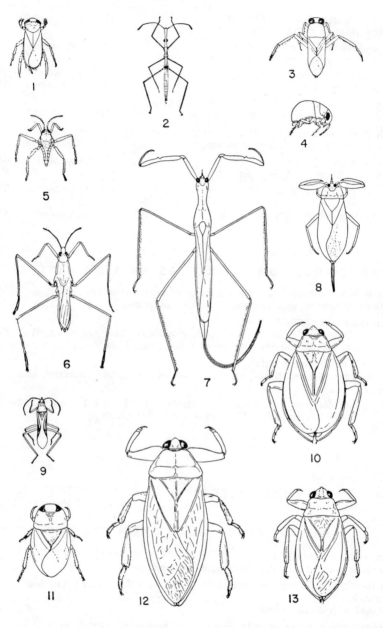

Figs.	Genus	Length	Distr.	Waters
10.	Abedus	25	G	static
13.	Belostoma	25	G	static
(12).	Benacus	37	G	static
(3).	Buenoa	6	G	static
6.	Gerris	15	G	both
1.	Hesperocorixa	6	G	static
2.	Hydrometra	10	G	static
12.	Lethocerus	37	G	static

Figs.	Genus	Length	Distr.	Waters
9.	Mesovelia	5	G	both
8.	Nepa	16	E	static
3.	Notonecta	12	G	static
11.	Pelocoris	9	E	static
4.	Plea	3	EG	static
7.	Ranatra	30	G	static
5.	Rhagovelia	4	G	lotic

5 –Apical appendages of the abdomen long and slender; tarsi 1-segmented:
NEPIDAE   *13*
–Apical appendages of the abdomen short and flat, retractile: BELOSTOMATIDAE
                                                                          14

6 –Head as long as entire thorax; both elongated. Both about 10 mm: HYDRO-
METRIDAE (1 genus)   *Hydrometra*   **XI-2**
–Head shorter than thorax including scutellum   *7*

7 –Claws of at least front tarsi distinctly anteapical, with terminal tarsal segment
more or less cleft   *8*
–Claws all apical, last tarsal segment entire   *9*

8 –Hind femur extending beyond apex of abdomen; intermediate and hind pairs
of legs approximated, very distant from front pair. Beak 4-segmented:
GERRIDAE   *17*
–Hind femur not extending much beyond apex of abdomen; intermediate pair
of legs about equidistant from front and hind pairs (except in Rhagovelia).
Beak 3-segmented: VELIDAE   *23*

9 –Antennæ 4-segmented. Membrane of wing without cells: MESOVELIIDAE. Tarsi
3-jointed (1 genus)   *Mesovelia*   **XI-9**
–Antennæ 5-segmented; tarsi 2-jointed: HEBRIDAE, length 2 mm   *Hebrus*

10 –Legs quite similar   *Plea*   **XI-4**
–Legs dissimilar, hind legs flattened and fringed for swimming   *11*

11 –Last segment of antennæ much shorter than penultimate   *Notonecta* **XI-3**
–Last segment of antennæ longer than the penultimate   *Buenoa* **XI-(3)**

12 –Front margin of prothorax deeply excavated for the reception of the head
                                                      Ambrysus
–Front margin of prothorax not deeply excavated for the reception of the head
                                                      Pelocoris   **XI-11**

13 –Body broadly oval and flat; legs not extremely long and slender; prothorax
much broader than the head; anterior femora but little longer than tibiæ
                                                          Nepa   **XI-8**
–Body elongate oval; legs not extremely long and slender; prothorax little
broader than head; anterior femora considerably longer than tibiæ *Curicta*
–Body very elongate; legs long and slender; prothorax narrower than head;
anterior femora considerably longer than tibiæ   *Ranatra* **XI-7**

14 –Mesothorax with strong midventral keel, membrane of hemelytra reduced
                                                        Abedus   **XI-10**
–Mesothorax without midventral keel, membrane of hemelytra not reduced *15*

15 –First or basal segment of the beak longer than the second; furrow of the
wing membrane nearly or quite straight. Length about 1 inch or less
                                                      Belostoma **XI-13**
–First segment of beak shorter than the second. Furrow of the wing membrane shallowly S-shaped. Length more than 1½ inches   *16*

16 –Anterior femora grooved for the reception of the tibiæ   *Lethocerus* **XI-12**
–Anterior femora not grooved for the reception of the tibiæ   *Benacus* **XI-(12)**

17 –Inner margin of the eyes arcuately sinuate behind the middle. Body comparatively long and narrow, abdomen long   *18*
–Inner margin of the eyes convexly rounded; body comparatively short and
broad, abdomen very short in some forms   *21*

18 –Pronotum sericeous, dull; antennæ comparatively short and stout   *19*
–Pronotum glabrous; shining; antennæ long and slender   *Tenagogonus*

19 –First segment of antennæ shorter than the second and third taken together *20*
–First segment of the antennæ longer than the second and third taken together
                                                          Gerris   **XI-6**

20 –Antennæ as long as half the body; sixth abdominal segment of the male
roundly emarginate   *Limnoporus*
–Antennæ not as long as half the entire length of the insect, not extending
beyond the thorax; sixth abdominal segment of male doubly emarginate
                                                          Gerris   **XI-6**

21 –First antennal segment much shorter than the other three taken together; not
much longer than the second and third taken together, and sometimes
shorter   22
–First antennal segment nearly equal to the remaining three taken together,
much longer than second and third; antennæ almost as long as entire body;
hind femora twice as long as hind tibia   *Metrobates*

22 –Fourth (apical) segment of antennæ longer than the third   *Trepobates*
–Fourth segment of antennæ never more than equal to third; basal segment
of anterior tarsi much shorter than second; hind femur equal to or much
shorter than hind tibia and tarsus taken together   *Rheumatobates*

23 –Last antennal segment longest   24
–First antennal segment longest   25

24 –Ocelli in contact with inner margin of the eyes   *Macrovelia*
–Ocelli absent   *Microvelia*

25 –Third segment of middle tarsus split and with feathery hairs set in the split
                                                            Rhagovelia

–Intermediate tarsi not split   *Velia*

REFERENCES: WATER BUGS (HEMIPTERA)

Usinger et al.
1956   *Aquatic insects of California,* University of California Press, Berkeley, 508 pp.
Hungerford
1948   *The Corixidae of the Western Hemisphere,* Univ. Kansas Sci. Bull., 32:827.

XII. CADDISFLY LARVAE (TRICHOPTERA)

KEY XII: CADDISFLY LARVAE

1 –Microcaddisfly larvæ; length usually less than 6 mm., abdomen usually
wider than thorax; no gills; some with a flattened case resembling a purse:
HYDROPTILIDAE (12 genera)   *Leucotrichia*   **XII-1**
–Large caddisfly larvæ; abdomen about same width as thorax; free-living or
case-making larvæ   2

2 –Caudal prolegs separate; no portable cases except in subfamily Glossosoma-
tinæ; remainder free-living or net-spinning   3
–Caudal prolegs fused so as to form an apparent tenth segment; larvæ construct a definite case   7

3 –Larval case turtle-shaped, its upper portion dome-shaped, the underside with
a flat strap   *Glossosoma*   **XII-2, 3**
–Larvæ are free-living or net-spinning; predacious; construct pupation cases
only   4

PLATE XII: CADDISFLY LARVAL HOUSES*

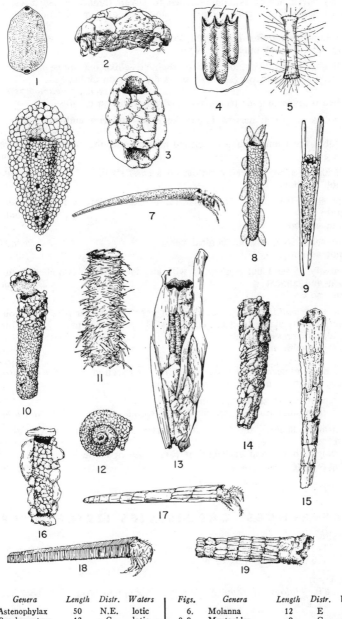

Figs.	Genera	Length	Distr.	Waters	Figs.	Genera	Length	Distr.	Waters
13.	Astenophylax	50	N.E.	lotic	6.	Molanna	12	E	static
18.	Brachycentrus	12	G	lotic	8, 9.	Mystacides	9	G	static
4.	Chimarra	8	G	lotic	16.	Neophylax	12	G	lotic
2, 3.	Glossosoma	9	G	lotic	19.	Phryganea	30	G	both
12.	Helicopsyche	6	G	lotic	11.	Platycentropus	24	E	lotic
10.	Hesperophylax	10	G	lotic	5.	Polycentropus	17	G	both
7.	Leptocella	11	G	both	15.	Ptilostomis	50	E	both
1.	Leucotrichia	6	G	both	17.	Triaenodes	8	G	static
14.	Limnephilus	18	G	both					

* For the worms themselves, see Plate VII-14, 15, 16.

4 –Free-living larvæ which remain attached to silken threads on surfaces; no rows of dense bushy gills; sclerotized shield on dorsum of ninth segment
Rhyacophila
–Net-forming larvæ living in a fixed location; usually live in swift-flowing water; net collapses outside the water **5**

5 –Larvæ live in a web-like tube back of their sieve-like net, larvæ are worm-like, very active and possess rows of bushy gills on abdomen
Hydropsyche, Cheumatopsyche **VII-15**
–Larvæ live in finger-like or trumpet-shaped tubes; no bushy gills **6**

6 –Larval tubes are long, narrow finger-like tubes, anterior end anchored
Chimarra, Dolophilodes **XII-4**
–Larval tubes are trumpet-shaped, opened at each end, the anterior end flared outward *Polycentropus* **XII-5**

7 –Cases made of sand, markedly similar to a snail shell *Helicopsyche* **XII-12**
–Cases not as above **8**

8 –Cases of sand, front and lateral margins broadened into a flat, shield-like structure *Molanna* **XII-6**
–Cases not as above **9**

9 –Cases of sand, long smooth conical cases *Leptocella* **XII-7**
–Cases not as above **10**

10 –Cases mostly of sand but sometimes with slender pieces of wood alongside; straight and conical *Mystacides* **XII-8, 9**
–Cases not so **11**

11 –Cases of sand, small stones, small shells or vegetable matter (the "log-cabin" cases); cylindrical or conical, usually slightly curved, usually round or triangular in cross-section LIMNEPHILIDAE **XII-10-14**
–Cases of vegetable matter **12**

12 –Cases of plant material which are square in cross section *Brachycentrus*
–Cases circular or triangular in cross section **13**

13 –Cases of plant material arranged in a spiral **14**
–Cases of plant material not arranged in a spiral *Ptilostomis* **XII-15**

14 –Cases, straight, slender and longer than larvæ, the posterior end of case is closed *Triaenodes* **XII-17**
–Cases straight, bulky and each end of case is open; larvæ in quiet or slow-flowing water *Phryganea* **XII-19**

REFERENCES: CADDISFLIES (TRICHOPTERA)

Betten, Cornelius
1934   *The caddisflies or Trichoptera of New York State.* N.Y. State Mus. Bull. 292, 576 pp., 61 text figs., 67 pls.

Denning, D. G.
1937   *The biology of some Minnesota Trichoptera.* Am. Ent. Soc. Trans. 63:17-43, 45 figs.

Kawamura and Tsuda
1951   *Trichoptera larvae,* reprinted from *Illustrated pocket book of insect larvae,* Tokyo.

Ross, H. H.
1944   *The caddisflies or Trichoptera of Illinois.* Ill. Nat. Hist. Sur. Bull., 23:1-326, 961 figs.

XIII. TWO-WINGED FLIES (DIPTERA)

Key to **Diptera** begins on page 56.

REFERENCES: TWO-WINGED FLIES (DIPTERA)

Alexander, C. P.
    1942   *The Diptera or true flies of Connecticut.* Fasc. 1. Conn. State Geol. and Nat. Hist. Survey Bull., 64:183-486.

Carpenter, S. J., and W. J. LaCasse
    1955   *Mosquitoes of North America (North of Mexico).* Univ. of Calif. Press, 360 pp.

Curran, C. H.
    1934   *The families and genera of North American Diptera.* Ballou Press, N.Y., 512 pp.

Johannsen, O. A.
    1934   *Aquatic Diptera.* Part I. Nemocera, exclusive of Chironomidae and Ceratopogonidae. Cornell Univ. Agr. Exp. Sta. Mem., 164:1-71.

    1935   *Aquatic Diptera.* Part II. Orthorrhapha-Brachycera and Cyclorrapha. Cornell Univ. Exp. Sta. Mem., 177:1-62.

    1937a   *Aquatic Diptera.* Part III. Chironomidae: Subfamilies Tanypodinae, Diamesinae, and Orthocladiinae. Cornell Univ. Agr. Exp. Sta. Mem., 205:1-84.

    1937b   *Aquatic Diptera.* Part IV. Chironomidae: Subfamily Chironominae, and Part V. Ceratopogonidae (by Lilian C. Thomsen). Cornell Univ. Agr. Exp. Sta. Mem., 210:1-80.

Malloch, John R.
    1915   *The Chironomidae, or midges, of Illinois, with particular reference to the species occurring in the Illinois River.* Bull. Illinois State Lab. Nat. Hist., 10:275-543.

Matheson, Robert
    1944   *Handbook of mosquitoes of North America,* 2nd ed., Comstock Press, Ithaca, N.Y.

Quate, Larry W.
    1955   *A revision of the Psychodidae (Diptera) in America north of Mexico.* Univ. Calif. Publs. Entomol. 10:103-273.

Wirth, W. E., and A. Stone
    1956   "Aquatic Diptera," in R. L. Usinger (ed.), *Aquatic insects of California.* Univ. of Calif. Press, Berkeley, Calif., pp. 372-482.

KEY XIII. CRANE FLY LARVAE

1 –Head free; tail long, partly retractile   2
 –Head telescoped or slipped back inside prothorax   3

2 –Color yellow or brown   *Liriope*   **XIII-24, 25**
 –Color rusty red or black   *Bittacomorpha*   **XIIIA-(24)**

3 –Body with numerous long spines longer than its own diameter   *Phalacrocera*   **XIIIA-29**
 –Body without such spines   4

4 –Spiracular disc at end of abdomen surrounded by 6 or 8 lobes   5
 –Spiracular disc at end of abdomen surrounded by 5 or fewer lobes   7

5 –Anal gills simple   6
 –Anal gills branched   *Aeshnasoma*   **XIIIA-27**

6 –Lobes of disc fringed with long hairs   *Holorusia*   **XIIIA-35**
 –Lobes of disc fringed with short hairs or none   *Tipula*   **XIIIA-7, 8**

7 –Body covered with flat leaf-like projections   *Triogma*   **XIIIA-26**
 –Body bare   8

8 –Body depressed; no lobes on spiracular disc   *Elliptera*   **XIIIA-5,6**
 –Body cylindric; lobes present on spiracular disc   9

9 –Spiracular disc with 5 lobes   10
 –Spiracular disc with less than 5 lobes   12

10 –Creeping welts beneath abdominal segments   *Helius*
 –No creeping welts present   11

11 –Color green   *Erioptera*   **XIIIA-12**
 –Color not green   *Trimicra*   **XIIIA-13**

12 –Spiracular disc with 4 lobes   13
 –Spiracular disc with 2 lobes   16

13 –Spiracular disc defined by ridge above; hair fringes short   *Hexatoma*   **XIIIA-23**
 –Spiracular disc not defined above by ridge; hair fringes long   14

14 –Disc bare across upper border   *Eriocera*   **XIIIA-20-22**
 –Disc hairy across upper border   15

15 –Lower border of spiracular disc cross lined with black   *Pilaria*   **XIIIA-9-11**
 –Lower border of disc little or not at all crossed with black   *Limnophila*   **XIIIA-1-3**

16 –Creeping welts black; spiracles invisible   *Antocha*   **XIIIA-17-19**
 –Creeping welts if present not black; spiracles easily visible   17

17 –Lobes long and densely clothed with long hairs   *Pilaria*   **XIIIA-9-11**
 –Lobes bare, or nearly so   18

18 –Lobes slender, tapering, body segments with transverse rows of bristles   *Rhapidolabis*   **XIIIA-28**
 –Lobes shorter, more blunt; body segments smooth   19

19 –Anal gills very unequally 2 segmented   *Pedicia*   **XIIIA-30-31**
 –Anal gills more equally divided into a larger number of segments   20

20 –Spiracles separated by their own width   *Tricyphona*   **XIIIA-4**
 –Spiracles separated by twice their own diameter   *Dicranota*   **XIIIA-14-16**

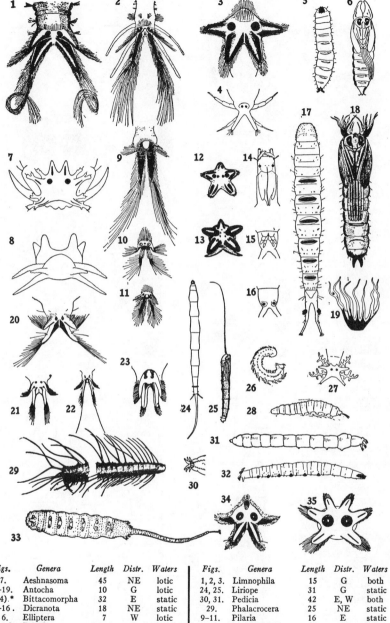

Figs.	Genera	Length	Distr.	Waters	Figs.	Genera	Length	Distr.	Waters
27.	Aeshnasoma	45	NE	lotic	1, 2, 3.	Limnophila	15	G	both
17–19.	Antocha	10	G	lotic	24, 25.	Liriope	31	G	static
(24). *	Bittacomorpha	32	E	static	30, 31.	Pedicia	42	E, W	both
14–16 .	Dicranota	18	NE	static	29.	Phalacrocera	25	NE	static
5, 6.	Elliptera	7	W	lotic	9–11.	Pilaria	16	E	static
20–2 2.	Eriocera	25–45	G	lotic	28.	Rhaphidolabis	9	G	lotic
12.	Erioptera	10	G	both	7, 8.	Tipula	16–55	G	both
33.	Eristalis	9	G	both	4.	Tricyphona	17	E	both
32, 34.	Helius	10	G	both	13.	Trimicra	15	G	lotic
23.	Hexatoma	14	G	both	26.	Triogma	19	E	?
35.	Holorusia	55	W	?					

Mostly from Dr. C. P. Alexander. * Differs by being rusty red in color.
Fig. 33 is a rat tailed maggot, (family Syrphidæ) larva of the Syrphus fly *Eristalis*.

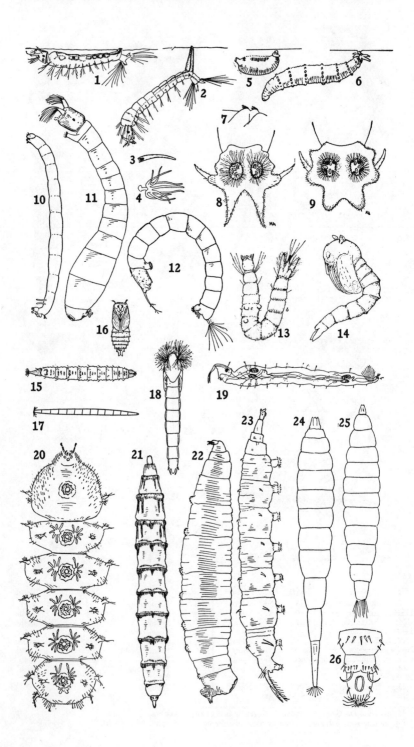

XIV. BEETLES (COLEOPTERA)

KEY XIVA: ADULTS OF COMMON AQUATIC BEETLES*

1 –First visible abdominal sternite completely divided at middle by hind coxal cavities (in HALIPLIDAE, bases of hind femora and first 2 or 3 sternites are hidden by expanded hind coxal plates, VII-4) *2*
  –First visible abdominal sternite not divided, extending uninterruptedly behind coxal cavities (if 2 pairs of eyes present, see GYRINIDAE) *7*

2 –Two pairs of eyes; middle and hind legs flat, tarsi of hind pair folding fanwise; length 3–16 mm. GYRINIDAE **VII-1**
  –One pair of eyes; middle and hind legs suited for crawling and swimming. hind pair never flat with folding tarsi *3* **XIVA-2**

3 –Hind coxæ expanded into large plates covering basal half of hind femora and first 2 or 3 abdominal sternites; Length 1.75–5.5 mm. HALIPLIDAE **VII-4**
  –Hind coxæ not expanded into large plates, not covering either basal half of femora or 2 or 3 sternites *4*

4 –Mid-metasternum with a small triangular sclerite just in front of hind coxæ; hind tarsi not flattened nor fringed with hairs, but simple, carabid-like; length 11—14.5 mm. AMPHIZOIDAE **XIVA-2**
  –Mid-metasternum without transverse suture delimiting a triangular sclerite in front of hind coxae; hind tarsi distinctly flattened, usually fringed with long hairs *5*

5 –Front tarsi clearly 5-segmented, fourth segment about as long as third; a line drawn from front of middle of prosternum to end of prosternal process is on an almost even plane *6*
  –Front tarsi apparently 4-segmented, true 4th segment minute, concealed in apex of third; prosternal process behind front coxæ at an angle to middle of prosternum in front of coxæ; length 1.2—6 mm. (subfamily HYDROPORINAE) DYTISCIDAE **VII-3**

6 –Scutellum fully visible as in XIVA-2, or if concealed then each hind tarsus has a single straight claw and tarsal segments are lobed behind on outer side; front tibiæ without a strong recurved spur at outer apical angle; length 4—35 mm. DYTISCIDAE **VII-3**

* Key to larvæ on page 61.

←——————————————————

Figs.	Genera	Length	Distr.	Waters	Figs.	Genera	Length	Distr.	Waters
23.	Atherix	16	G	lotic	25.	Odontomyia	12–50	G	static
1.	Anopheles	10	G	static	15, 16.	Psychoda	6	G	static
20.	Bibiocephala	10	W	lotic	22.	Sarcophaga	14	E	static
(20)a.	Blepharocera	8	E, W	lotic	5–8.	Sepedon	11	G	static
17.	Ceratopogon	5	G	static	11, 4.	Simulium	7	G	lotic
19.	Chaoborus	8	G	static	24.	Stratiomys	27–40	G	static
10, 18.	Chironomus	2–22	G	both	21.	Tabanus	20–55	G	static
(21)b.	Chrysops	12	G	static	(12)c.	Tanypus	6	G	both
2, 3.	Culex	7–12	G	static	12.	Tanytarsus	3–7	N	both
13, 14.	Dixa	6–10	G	static	9(5.6).	Tetanocera	11	G	static
(25), 26.	Euparyphus	12	G	lotic					

a Differs by lack of lateral pedunculate processes.
b Differs by having the last antennal joint longer than one preceding.
c Differs by lacking the peduncle under the antenna.

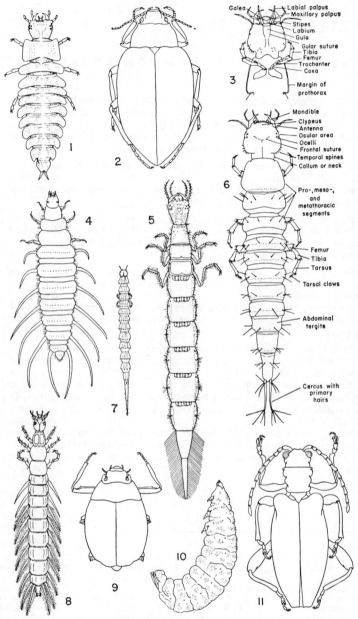

Labial palpus
Maxillary palpus
Galea
Stipes
Labium
Gula
Gular suture
Tibia
Femur
Trochanter
Coxa
3
Margin of prothorax

Mandible
Clypeus
Antenna
Ocular area
Ocelli
Frontal suture
Temporal spines
Collum or neck
6
Pro-, meso-, and metathoracic segments

Femur
Tibia
Tarsus
Tarsal claws

Abdominal tergite

Cercus with primary hairs

Figs.	Genera*	Length	Distr.	Waters	Figs.	Genera*	Length	Distr.	Waters
(6).	Agabus (l.)	10–15	G	both	10, 11.	Donacia (l. & a.)	10	G	static
1, 2.	Amphizoa (l. & a.)	12	W	lotic	3, 6.	Dytiscid (l.)	70	G	both
4.	Berosus (l.)	10	G	both	(8).	Gyrinus (l.)	6–14	G	both
(6).	Colymbetes (l.)	22	G	both	7.	Haliplus (l)	10	G	both
5.	Cybister (l.)	75	G	static	(6).	Ilybius (l.)	11–15	G	static
8, 9.	Dineutus (l. & a.)	11–28	E, C	both	(6).	Rhantus (l.)	10–12	G	both

* a. = adult; l. = larva.

–Scutellum hidden by bases of elytra and hind margin of pronotum XIVB-11; hind tarsi with 2 slender curved claws of equal length, sides of tarsi nearly parallel; front tibiæ with a strong recurved spur at outer apical angle (XIVB-11) (except in *Notomicrus,* length less than 1.5 mm., Louisiana and Florida); length 1.2—4.5 mm.   NOTERIDAE

7 –Antennæ short, true segment 6 cup-shaped (XIVB-14), segments beyond that forming a differentiated pubescent club, those before cupule simple and glabuous; maxillary palpi nearly always longer than antennæ; head usually with a Y-shaped impressed line on vertex   8
   –Antennæ not so constructed, and longer than maxillary palpi; head without a Y-shaped impressed line on vertex   9

8 –Antennal club of 5 pubescent segments; tiny beetles not over 2.5 mm. in length   HYDRAENIDAE   **XIVB-(13)**

   –Antennal club with only 3 pubescent segments beyond cupule; size 1.2—45 mm., but species approaching 1 mm. in length are convex and rounded   HYDROPHILIDAE   **VII-2**

9 –All tarsi actually 5-segmented but appearing to be 4-segmented; segment 4 very small and nearly concealed within lobes of third segment; or equally small and actually joined to base of segment 5; first 3 segments dilated, with hairy pads beneath; length 5—10 mm.   CHRYSOMELIDAE   **XIVA-11**
   –All tarsi with 5 or fewer segments, but not fitting above description   10

10 –Front coxæ more or less conically projecting; hind margin of prothorax never crenulate; maxillary palpi shorter than first 4 antennal segments combined; length 2.5—6 mm.   HELODIDAE
   –Front coxæ variously formed; if projecting, then hind margin of prothorax is crenulate, or maxillary palpi are more than half as long as antennæ   11

11 –Six or 7 abdominal sternites present; length 2—5.5 mm.   PSEPHENIDAE   **XIVB-19**
   –Five abdominal sternites   12

12 –Anterior coxæ transverse, with exposed trochantin; antennæ generally short and serrate; length 3—8 mm.   DRYOPIDAE   **VII-5**
   –Anterior coxæ globular, nearly always without trochantin; antennæ usually slender; length 1—8 mm.   ELMIDAE   **VII-5**

KEY XIVB: LARVAE OF COMMON AQUATIC BEETLES

1 –Legs apparently 5-segmented, tarsal claws 2, except in HALIPLIDAE which are 1-clawed   2
   –Legs apparently 4-segmented, tarsal claws 1   18

2 –Abdomen with 4 hooks at end and with lateral gills on all segments: GYRINIDAE   3
   –Abdomen not with 4 hooks at end; abdominal gills present or not   4

3 –Head subcircular, neck narrow and distinct   DINEUTUS
   –Head elongate, neck about as wide as rest of head   *Gyrinus, (Gyretes)*

4 –Abdomen with 9 or 10 segments: HALIPLIDAE   5   **VII-6, 7**
   –Abdomen with 8 visible segments   6

61

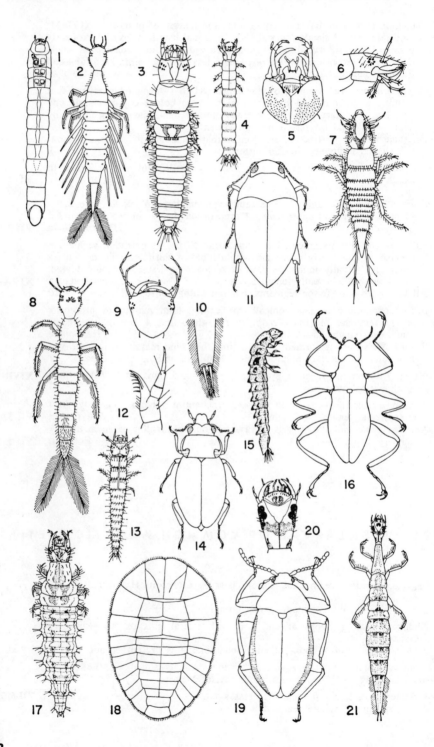

5 –Abdomen 9-segmented; body filaments segmented, each half as long as
body *Peltodytes* **VII-6**
–Abdomen 10-segmented; body filaments not segmented, not much longer
than length of one body segment *Halipus, (Brychus)* **VII-7**

6 –Body cylindrical, elaterid (wireworm-like), not over 12 mm. long; legs short,
suitable for burrowing; larvæ of most genera still unknown NOTERIDAE
–Body not cylindrical but in part broadened and flattened, often strongly
tapering; legs longer, suitable for walking and/or swimming *7*

7 –Body flattened, thoracic and abdominal sides greatly expanded into plates;
gular suture median, simple: AMPHIZOIDAE *Amphizoa* **XIVA-1**
–Sides of body not expanded, plate-like; gular suture double at least anteriorly
DYTISCIDAE *8*

8 –Head with a frontal projection or "snout" *9*
–Head without a frontal projection *10*

9 –Frontal projection with a notch at each side
(*Hygrotus, Deronectes, Oreodytes,* in part), *Hydroporus* **XIVB-6,**
7
–Frontal projection without notches (*Hydrovatus,* some *Oreodytes*), *Bidessus* **XIVB-(7)**

10 –Stipes of maxillary palpus broad, suboval, usually with 1 or 2 strong inner
marginal hooks *11*
–Maxillary stipes long, slender, without inner marginal hooks *17*

11 –Abdominal segments 7 or 8 or both with even fringe of long swimming hairs
16
–Abdominal segments 7 and 8 without specialized even fringe of stiff, swimming hairs *12*

12 –Fourth (apical) antennal segment more than ⅔ length of third
Rhantus, Colymbetes
–Fourth antennal segment less than ⅔ length of third *13*

13 –Mandibles toothed along inner edge *Copelatus*
–Mandibles not toothed along inner edge *14*

14 –Cerci usually with only 7 primary hairs in 2 whorls *15*
–Cerci with numerous secondary hairs *Laccophilus* **XIVB-8**

15 –Lateral margin of head compressed or keeled, with temporal spines on a line
which would intersect the ocelli or pass just below them *Ilybius*
–Lateral margin of head not keeled, temporal spines on a line which would
run well below ocelli *Agabus*

16 –Six anterior abdominal segments with lateral gills *Coptotomus* **XIVB-20,**
21
–No gills on abdominal segments (*Thermanectus*), (*Graphaderus*), *Acilius* **XIVB-**
(11)

Figs.	Genera*	Length	Distr.	Waters	Figs.	Genera*	Length	Distr.	Waters
(11).	Acilius (a.)	23	G	static	4.	Hydrochus (l.)	3–5	G	both
(7).	Bidessus (l.)	1.5–3.5	G	both	5.	Hydrophilus (l.)	35–45	G	both
2.	Coptotomus (l.)	7–12	G	static	6, 7.	Hydroporus (l.)	4–6	G	both
(3).	Cymbiodyta (l.)	6–10	G	both	(13).	Hydraena (l.)	3.5	G	both
1.	Dryopid (l.)	10–20	G	lotic	8.	Laccophilus (l.)	8	G	both
9, 10.	Dytiscus (l. & a.)	30–70	G	both	(13).	Limnebius (l.)	3	G	both
16.	Elmid (a.)	1.5–7.0	G	lotic	12, 13, 14.	Ochthebius (l. & a.)	4	G	both
3.	Enochrus (l.)	5–8	G	both	15.	Promoresia (l.)	4.5	E	lotic
(13).	Helophorus (l.)	5–10	G	both	18, 19.	Psephenus (l. & a.)	7	G	lotic
11.	Hydrocanthus (a.)	7	E, S	lotic	20, 21.	Thermonectus (l.)	16	G	static
(17).	Hydrochara (l.)	16–26	G	both	17.	Tropisternus (l.)	10–12	G	both

* a=adult; l=larvae

63

17 –Clypeus dentate   *Cybister*
   –Clypeus not dentate   *Hydaticus* and *Dytiscus*

18 –Cerci segmented, usually movable (examine end of abdomen carefully); cerci often very short, or retracted into a terminal breathing pocket in eighth abdominal segment, in Hydrophilidae   *19*
   –Cerci, if present, not segmented but solidly united at base   *29*

19 –Maxilla with 2nd and 3rd segments (= stipes and palpiger) separate, stipes without or with only rudimentary lacinia, palpiger usually carrying a small finger-like galea; mature larvæ 5 to 55 mm. long: HYDROPHILIDAE   *20*
   –Maxilla with stipes and palpiger fused, stipes with a large lacinia, palpiger without finger-like galea; tiny larvæ, up to 4 mm. long: HYDRAENIDAE
                              Limnebius, Hydraena, Ochthebius   **XIVB-12,**
                                                                  (13), 14

20 –Eight complete abdominal segments, ninth and tenth reduced and forming a stigmatic atrium (except in *Berosus* which has 7 pairs of very long abdominal gills)   *21*
   –Nine complete abdominal segments, tenth reduced but distinct   *Helophorus*   **XIVB-**
                                                                  (13)

21 –Antennæ inserted nearer lateral margins of head than are the mandibles
                                                        Hydrochus   **XIVB-4**
   –Antennæ inserted farther from lateral margins of head than are the mandibles
                                                              22

22 –Seven pairs of very long gills on sides of abdomen; abdominal segments 9 and 10 reduced but without stigmatic atrium   *Berosus*
   –Abdominal gills not nearly so prominent, or absent; breathing pocket present   *23*

23 –Femora without fringes of swimming hairs; mature larvæ smaller, 4 to 8 mm. long   *24*
   –Femora with fringes of long swimming hairs; mature larvæ larger, 15 to 50 mm. long   *27*

24 –Frontal sutures parallel   *Laccobius*
   –Frontal sutures not parallel   *25*

25 –Mandibles unlike, right with 2 teeth, left with 1; abdomen with prolegs on segments 3 to 7   *Enochrus*   **XIVB-3**
   –Mandibles alike, each with 2 or 3 inner teeth   *26*

26 –Labro-clypeus with 5 distinct teeth, outer left one a little distant from rest; mandibles with 3 inner teeth.   *Hydrobius*
   –Labro-clypeus with at least 6 teeth; mandibles with 2 inner teeth
                                  (*Helochares*), *Cymbiodyta*   **XIVB-(3)**

27 –Head subspherical; each mandible with 1 inner tooth   *Hydrophilus*   **XIVB-5**
   –Head subquadrangular; each mandible with 2 or more teeth   *28*

28 –Mentum with sides nearly straight; lateral abdominal gills rudimentary, indicated by tubercular projections each with several terminal setae
                                                      Tropisternus   **XIVB-17**
   –Mentum with sides converging in basal half; abdominal gills fairly well developed, pubescent   *Hydrochara*   **XIVB-17**

29 –Body broad, depressed; head not visible in dorsal view   PSEPHENIDAE   **XIVB-18**
   –Body usually slender, round or triangular in cross-section; head visible in dorsal view   *30*

30 –Ninth abdominal segment with a ventral movable operculum, closing a caudal chamber   *31*
   –Ninth abdominal segment without an operculum   *32*

31 –First 5 or 8 abdominal sternites each with median fold (3 genera)   DRYOPIDAE
   –Abdominal sternites flattened or evenly convex   ELMIDAE

32 –Abdomen with tufts of filamentous gills restricted to anal region and sometimes retracted into a caudal chamber (open and examine apex of abdomen), body form more elliptical and flattened; antennæ multiarticulate,
   cockroach-like, usually longer than head and thorax combined.
                                                              HELODIDAE
   –Without conspicuous abdominal gills; antennæ 3-segmented, short, broad,
   hardly projecting beyond outline of head; spiracles of eighth abdominal
   segment in part specialized into two, dorsal, projecting, spur-like processes:
   CHRYSOMELIDAE   *Neohaemonia, Donacia*

REFERENCES: BEETLES (COLEOPTERA)

Böving, A. G., and F. C. Craighead
   1930   *An illustrated synopsis of the principal larval forms of the order Coleoptera.*
          Entomologica Americana (N.S.) 11(1):1-80, (2) 81-160, incl. pls. 1-36 (Re-
          printed, 1953).

Böving, A. G., and K. L. Hendricksen
   1938   *The developmental stages of the Danish Hydrophilidae* (Insects-Coleoptera).
          Videnskabelige Meddelelser fra den Naturhistoriske Forening i Kobenhavn, 102:
          27-162, 55 text figs.

Dillon, E. S., and Lawrence S. Dillon
   1961   *A manual of common beetles of eastern North America*, 884 pp., Row, Peter-
          son & Co., Evanston, Ill.

Richmond, E. A.
   1920   *Studies on the biology of aquatic Hydrophilidae.* Bull. Amer. Mus. Nat. Hist.,
          42:1-94, pls. I-XVI.

Wilson, C. B.
   1923   *Water beetles in relation to pondfish culture with life histories of those found
          in fishponds at Fairport, Iowa.* Bull. Bur. Fish. (Washington, D.C.), 39:231-345,
          148 text figs.

XV. FRESH-WATER FISHES

Most of the common game fishes caught by anglers may be identified using the keys presented here. Aside from the main key (XVA) to the major groups that follows herewith, separate keys are given to Atlantic salmon, trout, and charrs (SALMONIDAE, Key XVB), perch (PERCIDAE, Key XVC), sunfishes and basses (CENTRARCHIDAE, Key XVD) and to adults and young of Pacific salmon (genus *Oncorhynchus,* Keys XVE and XVF).

The key gives specific as well as generic names where species may be easily separated, but more often refers to families only. With families having large numbers of genera and species such as in the minnow (CYPRINIDAE), sucker (CATOSTOMIDAE), catfish (ICTALURIDAE), whitefish (COREGONIDAE), and cottid (COTTIDAE) families, readers are referred to Hubbs and Lagler (1947) and Eddy (1957), the full references to which are listed at the bottom of the first page of the main key.

Strictly marine fishes are omitted. Of those marine forms that enter fresh water occasionally, only a few extremely common forms are given, such as the lampreys (PETROMYZONIDAE), sturgeon (ACIPENSERIDAE), eel (ANGUILLIDAE) and striped bass (SERRANIDAE). Also omitted are the blind cave fishes (AMBLYOPSIDAE), and many euryhaline forms that occur in estuary areas, such as the flatfishes, flounders, mullets, pipefishes, seahorses, gobies, anchovies, tarpon, needlefishes, ten-pounders, and characins.

Plate XVA presents the topography of a hypothetical spiny-rayed fish, indicating various measurements and structures commonly used in the identification of fishes. External and easily located characters are used for the most part.

DEFINITIONS AND COUNTS OF FISH STRUCTURES

Barbels: Fleshy "whiskers" found near the mouths of catfish (**XVB**-6), carp, sturgeon, and other fishes.

Branchiostegal rays: Bony rays that support the gill membranes on the lower side of the lower jaw (**XVA**-a). All are counted, and care should be exercised to count those most anterior in position that are often short and concealed.

Fins, paired: Pectorals and pelvics (**XVA**-a). Pelvic fins are often called ventral fins. These are said to be abdominal in position when they are located in the posterior half of the region between bases of the pectorals and the anus, as they are in the salmon and the trouts. They are called thoracic when located close to the pectorals in the anterior half of the body, as in sculpins, sunfishes, and basses. The *origin* of any given fin is its most anterior basal portion.

Fins, vertical: Dorsal, anal, and caudal (**XVA**-a, b). A small, fleshy fin lacking rays, the adipose fin, is often found in soft-rayed fishes (**XVA**-b) be-

PLATE XVA: FISH STRUCTURES

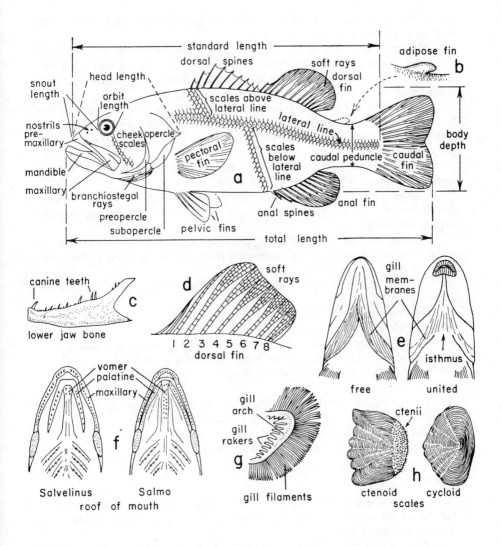

a. general features of a spiny-rayed fish
b. adipose fin showing its location on caudal peduncle when present
c. canine teeth on jawbone (dentary) of a yellow pikeperch (*Stizostedion*)
d. method of counting the number of principal dorsal fin rays in soft-rayed forms such as minnows, suckers, salmonids, and allied forms
e. shows the gill or branchiostegal membranes free (left) and united (right) to the isthmus on the underside of the head
f. comparison of the dentition in the roof of the mouth of a charr (*Sal-*

velinus) with a trout (*Salmo*). (Note that in *Salmo* the teeth extend down the shaft of the vomer while in *Salvelinus* they are restricted to the crest or head of this bone.)
g. position of the gill rakers and gill filaments on a single gill arch
h. main difference between cycloid and ctenoid scales. Fish with ctenoid scales such as bass, sunfish or perch feel rough in the hand while fish with cycloid scales such as trout or minnows feel slippery and are difficult to hold. The ctenii account for the difference.

tween the dorsal and caudal fins. Present in salmon, trout, smelt, whitefishes, and other primitive forms.

Fin ray counts: Generally only principal rays are enumerated (**XVA**-d), since rudimentary rays are hard to see or find. Spines are counted separately from soft rays. Probing with a needle is often necessary to locate extremely short spines located close to the anterior edges of fins.

Gill rakers: On the inner surfaces of the gill arches whose outer edges bear gill membranes. They may be long and slender or short and spine-like (**XVA**-g). The count is made on the first arch and includes the entire number present.

Intromittent organ: Modified anal fin of some male fishes to form a copulatory organ (**XVB**-12).

Lateral line: Easily seen when present on the side of fishes (**XVA**-a). It consists of a series of tubes and pores on the scales opening into sensory organs internally.

Maxillaries: The bones of the upper jaw. The first bone on each side of the midline of the upper jaw is the premaxillary. The second bone is the maxillary (**XVA**-a).

Papillae: Small, fleshy protuberances observed on the lips of suckers and other fishes.

Parr marks: Dark vertical bars seen on the sides of young trout, salmon, and other closely allied forms (**XVD**).

Pelvic process: An elongate appendage found at the base of the pelvic fin in some fishes. It is also called the axillary process or pelvic appendage (text fig. 3).

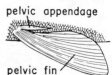

Pyloric caecae: Small, finger-like processes that encircle the intestine where it leaves the stomach. Only the number of tips are counted and this should be done prior to preservation for the sake of accuracy.

Fig. 3. Pelvic appendage

Scale counts:

1. Above lateral line (**XVA**-a): This is made from the origin (front) of the dorsal fin downward following a normal row to, but not including, the scale in the lateral line.

2. In lateral line: This is a count of the number of pores in the lateral line or, if no lateral line is present, of those in the normal position of a lateral line (**XVA**-a). Count begins with the first scale on the shoulder and ends with the last scale to touch, but not overlap for more than one half of its total length, the caudal peduncle axis. Scales on the base of the caudal fin are not included even if they are well developed and bear pores. The caudal axis can easily be determined by noting the crease formed by flexing the tail.

3. Below lateral line: Make this count upwards and forward from the origin of the anal fin to, but not including, the lateral line scale (**XVA**-a).

4. Cheek scales: This is the number of rows crossing an imaginary line from the eye to the angle of the preopercular bone (**XVA**-a).

Scutes: Hard bony or horny plates like those seen on sturgeon (**XVB**-2).

Teeth:

1. *Basibranchial:* These, when present, should not be confused with teeth on the tongue in the front of the floor of the mouth. They are found behind the tongue on the basibranchial bones at the bases of the gill arches. In cutthroat trout they usually occur between the first and second, and second and third, gill arches. Except in large, mature specimens, they are hard to see. Often mistakenly called hyoid teeth.

2. *Canine:* Strong teeth found on the jaws of pikeperch (*Stizostedion,* **XVA**-c) and many other predatory fishes.

3. *Palatine:* Located on the paired palatine bones in the roof of the mouth (**XVA**-f).

4. *Vomerine:* Teeth on the single vomerine bone in the middle of the roof of the mouth (**XVA**-f).

KEY XVA: FRESH-WATER FISHES*

1 –Mouth a sucking disc, without true jaws; gill openings more than four on each side: Lamprey family (PETROMYZONTIDAE)
Pacific lamprey, Entosphenus, et al
–Mouth with true jaws; a single gill opening, covered by an opercle on each side   *2*

2 –Body with rows of bony shields; snout conical; mouth on ventral side of head, preceded by four barbels; dorsal lobe of tail fin much longer than ventral lobe: Sturgeon family (ACIPENSERIDAE), 2 genera:
  –Caudal peduncle slender and entirely covered by bony shields; lower lip with 4 papillose lobes (text fig. 4a)   *Scaphirhynchus*
  –Caudal peduncle deep, covered by shields only on sides; lower lip with 2 papillose lobes (text fig. 4b)   *Acipenser*   **XVB-2**
–Body without shields or scales; snout long and paddle-like: Paddlefish family
  (POLYODONTIDAE)   paddlefish, *Polyodon spathula*   **XVB-3**
–Body with or without scales, snout not paddle-like   *3*

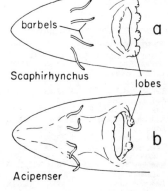

Fig. 4. Pappillose lobes on lower lips of two genera of sturgeon (after Eddy, 1957).

* Modified from Hubbs and Lagler, *Fishes of the Great Lakes Region,* Bull. 26, 1947, ranbrook Institute of Science, 186 pp., and Samuel Eddy, *How to Know the Freshwater shes,* 1957, Wm. C. Brown Co., Dubuque, Iowa, 253 pp.

3 –Body covered with thick, diamond-shaped scales; head produced into a strongly toothed beak: Gar family (LEPISOSTEIDAE)   gar, *Lepisosteus*   **XVB-4**
–Body without such scales; head not produced into beak   *4*

4 –Dorsal fin elongate, extending almost to base of tail fin, body not eel-shaped: Bowfin family (AMIIDAE)   bowfin, *Amia calva*   **XVB-5**
–Dorsal fin much shorter   *5*

5 –Dorsal fin with only one spine (carp, goldfish, and catfish) or without hard, spinous rays; pelvic fin without a spine   *6*
–Dorsal fin with more than one spine; pelvic fin with one or more spines   *20*

6 –Conspicuous barbels present above and below mouth (catfish; only above mouth in carp) or with single median barbel below jaw (cod family)   *7*
–Barbels absent around mouth   *8*

7 –Body scaleless; pectoral and dorsal fins each with hard, spinous ray: Catfish family (formerly AMEIURIDAE)   ICTALURIDAE   **XVB-6**
–Body scaled; no hard, spinous ray in pectoral or dorsal fins: Cod family (GADIDAE), 2 genera:
  –Dorsal fin divided into two parts, the second part very long, extending almost to tail fin; anal almost as long as dorsal fin
burbot, *Lota lota*   **XVB-7**
  –Dorsal fin divided into three parts, anal fin much shorter
tomcod, *Microgadus tomcod*   **XVC-16**

8 –Body eel-shaped, with true jaws, dorsal fin continuous with tail fin, pelvic fins absent: Fresh-water eel family (ANGUILLIDAE)
American eel, *Anguilla rostrata*
–Body not eel-shaped, dorsal fin not continuous with tail fin; pelvic fins present   *9*

9 –Tail fin rounded   *10*
–Tail fin not rounded   *12*

10 –Pelvic fins lying mostly behind origin of dorsal fin; premaxillaries not protractile; dark vertical bar at base of tail fin: Mudminnow family (UMBRIDAE)   mudminnow, *Umbra*   **XVB-9**
–Most (or all) of length of pelvic fins lying ahead of origin of dorsal fin; premaxillaries protractile   *11*

11 –Anal fins of male and female alike; third anal ray (counting rudiments) branched: Killifish family (CYPRINODONTIDAE)   *Fundulus* et al   **XVB-11**

Figs.	Common and Generic Names	Length (in.)	Distr.	Waters
5.	Bowfin or dogfish, *Amia*	15–30	S, C, E	both
7.	Burbot, *Lota*	20–30	C, N	both
6.	Catfish, *Ictalurus*	8–36	G	both
(15).	Charrs, *Salvelinus*[1]	6–36	G	both
8.	Eel, *Anguilla*	18–36	E	both
4.	Gar, *Lepisosteus*	18–96	S, C, N	both
16.	Grayling, *Thymallus*	12–20	N, W	both
11.	Killifish, *Fundulus*	3–6	S, E, C	both
1.	Lamprey, Pacific, *Entosphenus*	8–24	W	both
(1).	Lamprey, sea, *Petromyzon*	8–24	C, E	both
(1).	Lamprey, brook, *Lampetra*	6–9	E, C, W	both
(1).	Lamprey, silver, *Ichthyomyzon*	9–15	C, S, E	both

Figs.	Common and Generic Names	Length (in.)	Distr.	Water
12.	Mosquito fish, *Gambusia*	1–2	G	stati
9.	Mudminnow, *Umbra*	3–6	E, N	stati
3.	Paddlefish, *Polyodon*	24–72	C	both
14.	Pickerel and Muskelunge, *Esox*	12–40	S, E, C, N	both
(15).	Salmon, Atlantic, *Salmo*[1]	12–50	NE	both
(15).	Salmon, Pacific, *Oncorhynchus*[2]	6–48	W	both
13.	Shad, gizzard, *Dorosoma*	10–18	S, C, N	both
(13).	Shad, threadfin, *Dorosoma*	3–6	G	both
17.	Smelt, *Osmerus*	6–12	NE, NC	both
2.	Sturgeon, *Acipenser*	36–144	G	both
10.	Sucker, *Catostomus*	4–24	G	both
15.	Trout, *Salmo*[1]	6–36	G	both

[1] See Key XVB: Trout and Charrs.
[2] See Keys XVE and XVF: Pacific Salmon.

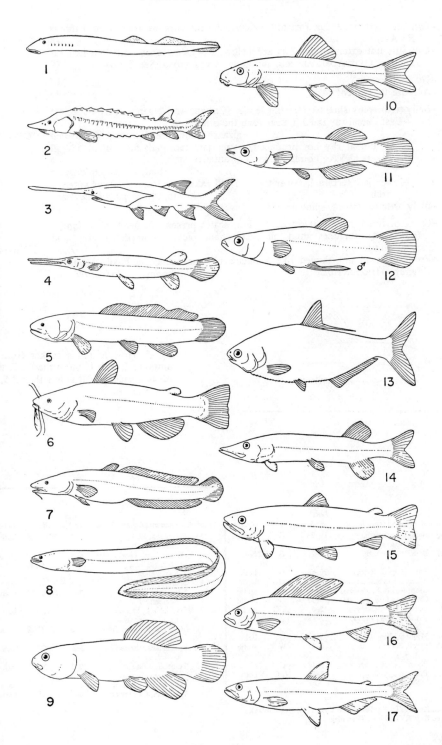

12 –Gill slits extended far forward below; gill membranes free from isthmus
(**XVA-e**)   *13*
–Gill slits not extended far forward below; gill membranes united to isthmus
(suckers, CATOSTOMIDAE; and minnows, CYPRINIDAE. See fig. 5)   *19*

13 –No adipose fin   *14*
–Adipose fin present   *16*

14 –Belly with spiny shields: Herring family (CLUPEIDAE), 2 genera and 4 species:
–Last dorsal ray much longer than the others
gizzard shad, threadfin shad, *Dorosoma*   **XVB-13**
–Last dorsal ray not produced; upper jaw terminal; more than 55 gill
rakers (**XVA-g**) on lower limb of anterior arch
shad, *Alosa sapidissima*
–lower jaw terminal, fewer than 55 gill rakers on lower limb of anterior
arch   alewife, *Alosa pseudoharengus*
–Belly without strong spiny shields   *15*

15 –Front of head shaped like a duck's bill; scales present on head: Pike family
(ESOCIDAE), includes mud pickerel, chain pickerel, northern pike and
muskellunge   *Esox*   **XVB-14**
–Front of head not shaped like a duck's bill; no scales on head: Mooneye
family (HIODONTIDAE)   mooneye, *Hiodon tergisus*

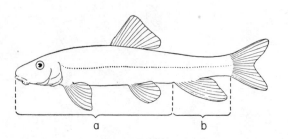

a   b

Fig. 5. How to tell a sucker from a
minnow: If a/b is more than 2.5, the
fish is a sucker; if it is less than 2.5, it
is a minnow.

Figs.	Common and Generic Names	Length (in.)	Distr.	Waters
10.	Bass, largemouth, *Micropterus*[1]	10–25	G	static
(10).	Bass, smallmouth, *Micropterus*	8–20	G	both
14.	Bass, striped, *Roccus*	20–48	G	both
13.	Bluegill, sunfish, *Lepomis*[1]	4–8	G	both
12.	Crappie, black, *Pomoxis*[1]	6–12	G	static
(12).	Crappie, white, *Pomoxis*	6–12	G	static
9.	Darter, *Poecilichthys*	2–6	E, C	both
4.	Drum or sheepshead, *Aplodinotus*	12–48	C	both
3.	Minnow, *Notropis*	3–5	G	both
(12).	Perch, Sacramento, *Archoplites*[1]	4–15	W	static
11.	Perch, Yellow, *Perca*[2]	6–14	G	both
15.	Pikeperch, yellow, *Stizostedion*	12–36	N, C	both

Figs.	Common and Generic Names	Length (in.)	Distr.	Waters
(15).	Pikeperch, blue, *Stizostedion*[2]	12–36	N, C, E	both
5.	Pirateperch, *Aphredoderus*	3–5	C, E	static
(15).	Sauger, eastern, *Stizostedion*[2]	10–15	N, C, E	both
6.	Sculpin, *Cottus*	2–5	G	both
8.	Silverside, brook, *Labidesthes*	2–3	N, C, SE	both
7.	Stickleback, brook, *Eucalia*	2–3	N, C	both
(7).	Stickleback, ninespine, *Pungitius*	2–3	N, C, E	both
(7).	Stickleback, threespine *Gasterosteus*	2–3	G	both
16.	Tomcod, Atlantic, *Microgadus*	8–12	NE	lotic
2.	Troutperch, *Columbia*	4–8	N	both
1.	Whitefish, *Coregonus*	8–24	G	both

[1] See Key XVD: Sunfishes and basses (Centrarchidae).
[2] See Key XVE: Perch (Percidae).

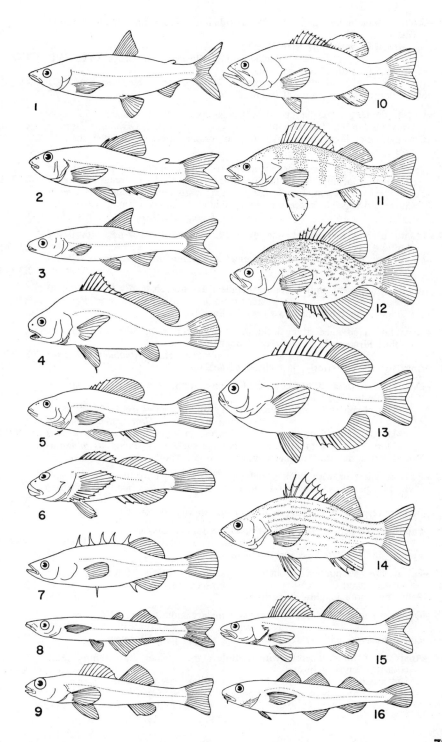

16 –Scales small, more than 100 in lateral line: Trout and salmon family
(SALMONIDAE):
  –Anal fin with more than 13 rays
              Pacific salmon, *Oncorhynchus* Keys XVE, XVF   **XVD**
  –Anal fin with 12 or fewer rays
       Atlantic salmon, trout, *Salmo, Charrs, Salvelinus* Key XVB   **XVB-15**
–Scales larger, fewer than 100 in lateral line   *17*

17 –Dorsal fin large, sail-like, bearing orange spots; with more than 15 rays:
Grayling family (THYMALLIDAE)   grayling, *Thymallus arcticus*   **XVB-16**
  –Dorsal fin not large or sail-like, without orange spots; with fewer than 15
rays   *18*

18 –Teeth strong; upper jaw bones extending behind middle of eye; pelvic process
(text fig. 3) lacking: Smelt family (OSMERIDAE)   smelt, *Osmerus*   **XVB-17**
  –Teeth weak or absent, jaw bones not extending behind middle of eye; a
short, fleshy pelvic process present: Whitefish family (COREGONIDAE)
        whitefishes, *Coregonus, Leucicthys, Prosopium*   **XVC-1**

19 –Mouth usually on ventral side of head, adapted to sucking, with fleshy
papillose lips: Sucker family (CATOSTOMIDAE)   sucker, *Catostomus*   **XVB-10**
  –Mouth usually terminal and not adapted to sucking: Minnow family (CYRPI-
NIDAE)   *Notropis*   **XVC-3**

20 –Adipose fin present; only two dorsal spines: Troutperch family (PERCOPSIDAE)
       western form, *Columbia,* eastern form, *Percopis*   **XVC-2**
  –No adipose fin; more than two dorsal spines   *21*

21 –Anus far in advance of anal fin, located near throat (indicated by arrow
on fig): Pirateperch family (APHREDODERIDAE)
        pirate perch, *Aphredoderus sayanus*   **XVC-5**
  –Anus located posteriorly, just ahead of anal fin   *22*

22 –Dorsal spines isolated, not connected to one another by membranes: Stickle-
back family (GASTEROSTEIDAE):
  –Dorsal spines 3, not including a very small spine at front of soft dorsal
        threespine stickleback, *Gasterosteus aculeatus*
  –Dorsal spines 5 or 6   brook stickleback, *Eucalia inconstans*   **XVC-7**
  –Dorsal spines 8 to 11,   ninespine stickleback, *Pungitius pungitius*
  –Dorsal spines not isolated, but connected to one another by membranes   *23*

23 –Body scaleless or with prickles only; pelvic fin with only three or four soft
rays; length of fish usually less than four inches: Sculpin family (COTTIDAE)
        sculpins, *Cottus*   **XVC-6**
  –Body covered with ordinary scales; pelvic fin with five soft rays   *24*

24 –Lateral line extended across tail fin: Drum family (SCIAENIDAE)
       fresh-water drum, *Aplodinotus grunniens*   **XVC-4**
  –Lateral line not extended across tail fin   *25*

25 –Pelvic fins abdominal, located far behind pectorals: Silverside family
(ATHERINIDAE)   brook silverside, *Labidesthes sicculus*   **XVC-8**
  –Pelvic fins thoracic, close to pectorals   *26*

26 –Dorsal fin single (but almost separated in largemouth bass, **XVC-**10)
        sunfish family, CENTRARCHIDAE, Key XVD
  –Dorsal fins two, entirely separate or only slightly joined together   *27*

27 –Anal spines three; opercle with well-developed spine: Bass family (SER-
RANIDAE): single genus   *Roccus*   **XVC-14**
    (4 species: striped bass, white bass, yellow bass and white perch)
  –Anal spines one or two; opercle without well-developed spine
        perch family, PERCIDAE, Key XVC

KEY XVB: TROUT (SALMO) AND CHARRS (SALVELINUS)

1 –Teeth absent on shaft of vomer (**XVA**-f); body with light spots on a darker background; scales very fine and imbedded, almost invisible to the naked eye: Charrs (*Salvelinus*)   2
    –Teeth present on shaft of vomer; body with dark spots on a lighter background, scales larger: Trouts (*Salmo*)   3   **XVB**-15

2 –Tail fin deeply forked; no wormlike mottling on dorsal part of body; found only in deep, cold lakes   lake trout, *Salvelinus namaycush*
    –Tail fin not deeply forked; mottled wormlike markings present on dorsal part of body; pronounced white borders on pectoral and pelvic fins
                                    eastern brook trout, *Salvelinus fontinalis*
    –Tail fin slightly forked, no mottled wormlike markings on dorsal part of body, but pale spots present on sides; found mostly in Northwest; never east of the Mississippi   Dolly Varden trout, *Salvelinus malma*

3 –Basibranchial teeth present, very tiny and difficult to see, located posterior to the tongue; generally a prominent red streak under each side of lower jaw; restricted to states west of the Mississippi
                                    cutthroat trout, *Salmo clarkii*
    –Basibranchial teeth absent, no red streak under jaw   4

4 –Adults with X-shaped spots on side; young with very prominent vertical bars (parr marks) on side; anal rays generally 9; anadromous in northern Maine and Canada; also landlocked in lakes of northern New England and Canada
                                    Atlantic salmon, *Salmo salar*
    –Adults with round spots on side; anal rays generally 10 to 12   5

5 –Tail fin without spots, or spot restricted to dorsal portion; often red, orange, or brown spots on side; no red lateral band; pelvic fins not white-tipped, general color yellowish   European brown trout, *Salmo trutta*
    –Tail fin profusely spotted, always black, never red, orange, or brown; pelvic fins white-tipped; general color silvery or pink; usually a red or pink lateral band   rainbow trout, *Salmo gairdnerii*
    –Tail with or without spots; belly bright orange or gold; orange or yellow-red band along lateral line, fading dorsally; body with strong parr marks
                                    golden trout, *Salmo aguabonita*

KEY XVC: PERCH (PERCIDAE)

1 –Preopercle strongly serrate; branchiostegals 7; fishes of medium to large size, maturing at a length of more than 6 inches   2
    –Preopercle with smooth, non-serrate edge; branchiostegals usually 6; small fishes, usually 1 to 4 inches in length   darters, *Poecilichthys* et al.   **XVC**-9

2 –No canine teeth; body with definite vertical dark crossbands
                                    yellow perch, *Perca flavescens*   **XVC**-11
    –Canine teeth strong (**XVA**-c); crossbands indefinite or lacking
                            sauger, yellow pikeperch, blue pikeperch, *Stizostedion*   **XVC**-15

KEY XVD: SUNFISHES AND BASSES
(CENTRARCHIDAE)

1 –Anal spines 3   *2*
–Anal spines 5 or more   *9*

2 –Scales small, 58 or more in lateral line, black basses:
  –Upper jaw, with mouth closed, not extending to hind margin of eye;
  14 to 18 rows of cheek scales
                                smallmouth bass, *Micropterus dolomieui*
  –Upper jaw extending beyond hind margin of eye; cheek scales in 9 to 12
  rows   largemouth bass, *Micropterus salmoides*   **XVC-10**
  –Scales large, 53 or fewer in lateral line   *3*

3 Teeth present on tongue; supramaxillary well developed, its length greater
  than the greatest width of maxillary   warmouth, *Chaenobryttus gulosus*
  –No teeth on tongue; supramaxillary reduced or absent (sunfishes)   *4*

4 –Pectoral fins short and rounded, contained about four times in length of
  fish   *5*
  –Pectoral fins long and pointed, longest rays near dorsal side of fin, contained
  about three times in length of fish   *6*

5 –Scales small, 44 or more in lateral line, opercle stiff to margin, its membrane
  wide and light colored; gill rakers long and slender
                                green sunfish, *Lepomis cyanellus*
  –Scales larger, 39 or fewer in lateral line; opercle flexible posteriorly, its
  membrane with a narrow red margin behind and usually below; gill rakers
  reduced and knoblike   longear sunfish, *Lepomis megalotis*

6 –Opercular bone stiff behind, gill rakers short and stout   *7*
  –Opercular bone flexible posteriorly; gill rakers long and slender   *8*

7 –Opercle with definite scarlet spot, cheeks with prominent blue and orange
  stripes in life   pumpkinseed, *Lepomis gibbosus*
  –Opercle with broad scarlet margin; cheeks without conspicuous blue and
  orange stripes   shellcraker sunfish, *Lepomis microlophus*

8 –Opercle bone extending almost to margin of "ear flap"; soft rays of anal
  usually 10 to 12   common bluegill sunfish, *Lepomis macrochirus*   **XVC-13**
  –Opercular bone extending only to middle of "ear flap"; soft rays of anal 7
  to 9   orangespotted sunfish, *Lepomis humilis*

9 –Dorsal spines 11 to 12; anal-fin base about one-half length of dorsal base
                                northern rock bass, *Ambloplites rupestris*
  –Dorsal spines 6 to 8; anal-fin base about equal to length of dorsal base,
  crappie   *10*

10 –Dorsal spines normally 7 to 8; body with dark irregular blotches
                                black crappie, *Pomoxis nigromaculatus*   **XVC-12**
  –Dorsal spines normally 6; body with definite dark, vertical bars, strongest
  dorsally   white crappie, *Pomoxis annularis*

KEY XVE: ADULT PACIFIC SALMON
(ONCORHYNCHUS)

1 –Gill rakers 30 to 50, long: red, sockeye, or blueback salmon
Oncorhynchus nerka
–Gill rakers 18 to 33, short   *2*

2 –Scales very small, 170 to 232 one row above lateral line: humpback or pink
salmon   *Oncorhynchus gorbuscha*
–Scales larger, less than 160 one row above lateral line   *3*

3 –Scales usually 19 to 26 above, and usually 15 to 24 below the lateral line.
No distinct black spots on back and caudal fin: chum or dog salmon
Oncorhynchus keta
–Scales usually 25 to 31 above, and usually 23 to 34 below the lateral line;
distinct black spots on back and caudal fin   *4*

4 –Anal rays usually 15 to 17; pyloric caeca 140 to 150; entire mouth black:
tyee, spring, chinook, quinnat, or king salmon *Oncorhynchus tshawytscha*
–Anal rays usually 12 to 15; pyloric caeca 50 to 83; gums whitish: coho or
silver salmon   *Oncorhynchus kisutch*

KEY XVF: YOUNG PACIFIC SALMON
(ONCORHYNCHUS)*

1 –Parr marks present   *2*
–Parr marks absent or at best very indistinct   Pink Salmon   **XVD**

2 –Parr marks largely above lateral line   *3*
–Parr marks extending below lateral line   *4*

3 –Parr marks faint, not rounded   Chum Salmon   **XVD**
–Parr marks small, often oval or rounded   Sockeye Salmon **XVD**

4 –Parr marks narrow, widely spaced, long   Silver Salmon   **XVD**
–Parr marks wide, closely spaced, rounded   King Salmon   **XVD**

* Less than eight inches in length.

PLATE XVD: YOUNG PACIFIC SALMON

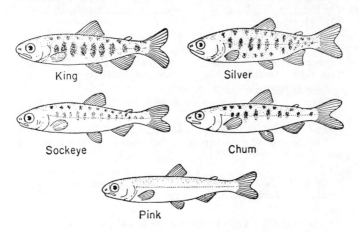

King   Silver

Sockeye   Chum

Pink

HOW TO TELL PACIFIC SALMON FROM TROUT

Aside from the keys given above, adult Pacific salmon (*Oncorhynchus*) may be separated easily and quickly from rainbow or steelhead and brown trout (*Salmo*) while fishing by the following characteristics:

1. Trout always have 9 to 12 rays in the anal fin while Pacific salmon have 13 or more;
2. In adult Pacific salmon the lining of the mouth is all black or mottled in blackish patches, while in trout it is always white;
3. In trout the branchiostegal rays are usually 10 to 12 while in the Pacific salmon these are usually more than 13.

REFERENCES: FISHES

Berg, Leo S.
1940   *Classification of fishes both recent and fossil.* Trav. Inst. Zool. Acad. Sci. U.S.S.R. Tome 5, Livr. 2: pp. 87-517.

Brown, Margaret E. (ed.)
1957   *The physiology of fishes.* 2 Vols. Vol. I Metabolism, 447 pp., Vol. II, Behavior, 526 pp. Academic Press, Inc. N.Y.

Carl, G. Clifford, and W. A. Clemens and C. C. Lindsey
1959   *The fresh-water fishes of British Columbia.* Brit. Col. Provincial Mus., Dept. Educa., Handbook No. 5 (3rd ed.), 192 pp.

Clemens, W. A., and G. V. Wilby
1961   *Fishes of the Pacific Coast of Canada.* Fish. Res. Bd. Canada, Bull. 68 (2nd ed.). 443 pp.

Eddy, Samuel
1957   *How to know the freshwater fishes.* Wm. C. Brown Co., Dubuque, Iowa, 253 pp.

Herald, Earl
1961   *Living fishes of the world.* Doubleday & Co., Garden City, N.Y., 304 pp.

Hubbs, Carl L., and Karl F. Lagler
1947   *Fishes of the Great Lakes Region,* Bull. 26, Cranbrook Institute of Science, 186 pp.

Jordan, David S., and B. W. Evermann
1896–1900 *The fishes of North and Middle America.* Bull. No. 47, U.S. Nat. Mus., 4 parts, 3313 pp.

Jordan, David S., and B. W. Evermann and H. W. Clark
1930   *Check list of the fishes and fishlike vertebrates of North and Middle America north of the northern boundary of Venezuela and Colombia.* Rept. U.S. Comm. Fish. for the fiscal year 1928. Reprinted 1955, 670 pp.

Lagler, Karl L., John E. Bardach, and R. R. Miller
1962   *Ichthyology.* 545 pp. John Wiley & Sons, Inc., New York, N.Y.

Nikol'skii, G. V.
1961   *Special ichthyology.* 538 pp. National Science Foundation, Washington, D.C., and the Smithsonian Institution, by the Israel Program for Scientific Translations, Jerusalem.

Schultz, Leonard P.
1936   *Keys to the fishes of Washington, Oregon and closely adjoining regions.* Univ. Wash. Publ. Zool. Vol. 2, No. 4: pp. 103-228.

Trautman, Milton B.
1957   *The fishes of Ohio.* Ohio State Univ. Press, 683 pp.

Part Two

METHODS OF SAMPLING AND

ANALYZING AQUATIC ORGANISMS

AND THEIR ENVIRONMENTS

THIS SECTION presents information on methods and equipment used in collecting aquatic organisms, along with suggestions for their care and preservation. Because of the great emphasis now being given to quantitative studies of both lotic and lentic environments, methods of taking quantitative stream and lake bottom samples are described. Methods of making chemical analyses of water are also presented. Lastly, a series of both general and specific studies are suggested as an aid to instruction.

Field trips to collect the organisms and to analyze the characteristics of each type of aquatic environment selected for study will form the basic core of training for students. These must be followed by indoor, laboratory periods devoted to analyses, identification, and write-up of the materials collected.

EQUIPMENT

1. Personal equipment
    a. Clothing suitable for field work including rubber boots, waders, and other waterproof gear.
    b. Notebook, hard and soft pencils for labelling and drawings or sketches.
    c. A pocket lens and dissecting kit of the type generally used in courses in biology.
    d. Containers suitable for bringing materials collected back to the laboratory in fit condition for study. Quart glass jars for living materials and vials of 70 per cent alcohol for "pickled" items are recommended.

2. Equipment for common use
    For the discovery and manipulation of the commoner invertebrate animals dwelling in the bottoms of both streams and lakes the following items are recommended:
    a. Ordinary white enamelled steel vegetable dishes about 7 × 10 × 2 inches are excellent for examining a fresh catch. When a dip net is emptied into the dish, the animals at once swim or crawl out from the trash and are easily seen against the white background. They can be picked up easily and uninjured on the lifter shown in figure 6.

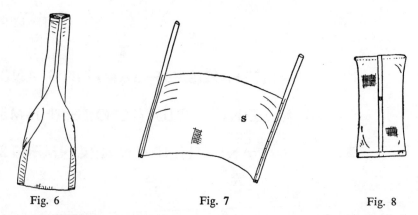

Fig. 6   Fig. 7   Fig. 8

Fig. 6. A lifter, made by folding the edges of a rectangular strip of wire cloth, say 3 × 6 inches. The lower edge should be a woven edge. A tinner's shears and 4-inch folding tongs are the tools needed for making this and the two next following pieces of apparatus.

Fig. 7. Hand screen for stream collecting. S is a sheet of wire screen 1 × 2 feet, woven edge below. The opposite edge, if not also woven, should be folded neatly, leaving no projecting wire-ends to prick fingers. The folded ends of the cloth are inserted in slots sawed in the handles and nailed fast there.

Fig. 8. Wire cage made from a square of wire cloth by first folding edges together to form a cylinder, and then cross folding the ends. It opens easily and closes securely with the fingers. Top edges should be woven to prevent pricked fingers.

b. A hand screen such as is shown in figure 7 is the most useful single tool for collecting small animals from rapid streams. It is held in the current by one person while another turns stones on the upstream side. The dislodged insects are washed by the current upon the screen, which is then lifted from the water. The larger and more conspicuous of them may be picked off by hand; but it is better to dump the catch into a white lined dish, as described above, and then to use a lifter. All of them, big and little, may be found in this way.

If the handles are made of light wood, such as willow or pine, the dumping is easy. Bring the two handles together and grasp them with one hand, then dash them downward against the other hand held stiffly over the dish of water. If aimed right the catch will all be discharged by the jolt into the dish.

c. If material is to be kept alive, there is no better retainer for it than the pillow cage shown in figure 8. It may be made of any desired mesh or dimensions. It may be immersed in stream or pond or tank. Half immersed it makes a good rearing cage for aquatic insects. A square yard of wire cloth makes four cages of suitable size.

d. Dip nets like those sold by dealers in collector's supplies are useful for light work about weed beds in ponds.

Fig. 9   Fig. 10

Fig. 9. Sieve net with detachable handle. As shown, it is supplied with a short handle and tow-line, ready to be used as a dredge. Sifting is done before lifting from water. Weight the bottom of the net for dredging.

Fig. 10. The apron net.

e. A sieve net of metal, well braced and strong, is most desirable for collecting bottom forms. It gathers the mud and sifts it at one operation. It is supplied with a long handle and is used as a rake from the shore (fig. 9). Where there is much loose trash on the bottom a common garden rake will answer some of the same purposes.

f. A very satisfactory dredge for use in any depth of water may be made out of the sieve net if a short wooden handle is substituted for the long one, a cord attached to its tip and the net lowered and dragged from a slow moving boat. The bottom stuff when brought to the surface is sifted before removal.

g. Plankton nets of fine silk bolting cloth are needed for obtaining the microscopic life of the open water. These are obtainable from dealers in biological supplies. No. 12 silk is of the mesh best suited to gathering a good catch quickly and in considerable variety.

h. For drawing water weeds ashore many kinds of weed hooks have been devised, but we have found nothing better than the weighted ring of barbed wire. A few other standard tools, such as hay knife and marl sampler for bog studies, are useful aids.

i. An apron net, figure 10, is a most useful collecting device and serves many purposes. It is so shaped at the front that it may be pushed through water weeds or under bottom trash. Its wide-meshed cover allows the animals to enter while keeping out the weeds and coarser trash. A final push through the water lands the catch at the rear, where it is easily accessible for picking-over by hand.

The smaller animals that are mixed with the trash in the net may best be found by dumping the contents of the net into a white dish, where they will at once reveal their presence by their activity. They may be taken from the water most easily and without injury on a lifter such as is shown in figure 6.

This net may also be used for scraping up and sifting the bottom mud and sand to obtain burrowers. It may be used for collecting the insects and other animals that abound among the loose stones in rapid streams by setting it edgewise against the bottom facing upstream and stirring the stones above it. The animals, dislodged by the stirring, will be swept by the current into the net.

81

Old leaf drifts, caught on obstructions in the current, may be stirred in the same way to get the animals hiding in them; but more stirring and overturning of the leaves will be necessary to dislodge them.

For a field class of twenty students, the following numbers of each item described above would be required:

white enamelled pans	10	handscreens	5
lifters	20	weed rings	2
dip nets	10	plankton nets	1
sieve nets	2	hand thermometers	1

CARING FOR FIELD COLLECTIONS*

Quart jars of water containing living organisms collected in the field should be left open and preferably kept in an icebox or cold room until they can be worked over. Specimen jars should not be left tightly sealed, as sealing will cause oxygen depletion in the jar which, coupled with warming of the water, will kill the organisms. It is far better to preserve the specimens in 70 per cent alcohol at the time of collection if a considerable period of time is to elapse between collection and study.

Fishes that the student wishes to keep alive should be placed in open-topped battery jars, aquaria, or other containers. Care should be taken to avoid overcrowding. A safe stocking rate is two or three suckers or minnows four inches long per gallon of water, but the number may be increased if the water can be aerated to maintain the oxygen supply. If your home water supply is free of chlorine, water dripped from a faucet through a wire screen placed over the top of the container will help to keep alive field collections of stream fishes as well as other organisms. On the other hand, pond dwelling animals from warm water need only to be kept in open containers without covers or lids.

Large, predacious dragonfly nymphs, diving beetle larvae, or other carnivorous forms should be held in separate containers or they will kill and eat soft-bodied forms such as mayfly or stonefly nymphs, dipterous larvae, damselfly nymphs, and shrimp. Similarly, small fishes should be separated from the larger ones.

All stones, rocks, mollusc shells, or other hard-bodied items should be removed from samples if they are to be transported any distance. Such articles rolling around the bottom of collection jars will grind up and destroy soft-bodied insect nymphs and larvae. On the other hand, a few leaves or pieces of aquatic plants placed in a jar will provide resting places for many of the animals collected and aid in getting them back alive.

* For methods of caring and rearing aquatic animals, consult *Culture Methods for Invertebrate Animals,* by J. G. Needham et al., Comstock Publishing Co., Ithaca, N.Y., 1937.

METHODS OF TAKING QUANTITATIVE
BOTTOM SAMPLES

Two types of equipment are most commonly used, the Surber square-foot stream bottom sampler (fig. 11), and the Ekman dredge (fig. 12). The latter secures samples from one-quarter square foot (36 square inches) while the former covers one square foot. The Surber sampler is designed for use in flowing waters of shallow streams and cannot be used satisfactorily in fast water over 18 inches in depth or in bottoms where the substrate is composed of large rubble and boulders. Both types of samplers may be purchased from biological supply houses.

Above: Fig. 11. Surber square-foot stream
bottom sampler.

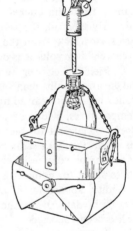

Right: Fig. 12. Ekman dredge for sampling
lake or pond bottoms.

Stream bottom sampling method. The Surber sampler is made of light brass and folds up compactly for carrying in the field. It consists of two frames hinged together. One frame carries the net that is stretched downstream by the flow of water for trapping bottom-dwelling forms, while the other marks the boundary of the square foot from which the collection is to be made. In use, the net is placed in the stream with the opening upstream. The frame marking the bottom area to be sampled should be worked carefully into the bottom. The operator holds it in position in the current by the pressure of one or both knees while using the hands to pick out the larger stones and wash organisms from them into the net. After the larger stones have been picked up, washed, and discarded, the remainder of the area inside the frame is gently churned up with the fingers to secure burrowing and other forms in the substrate. The contents of the net are then washed off by reversing it in a bucket about half full of water. Usually a few stones, with some sand and fine gravel, will be taken with the organisms. The latter may be easily removed from the trash by first whirling the water with the hand and then, while the water is turning rapidly, decanting through a 28 mesh soil sieve. The organisms, being lighter than the debris, will float up and can be concentrated in the sieve. Do not pour sand, gravel or small stones from the bucket into the sieve. Add more water to the bucket, whirl, and repeat the process until all the organisms have been removed from the debris. To make sure none are present, dump the trash into a white enamel pan and check it. If the stirring and decanting into the sieve has been effective, none will be found.

The next step is to transfer the sample from the sieve into a bottle. If the organisms, with accompanying light trash, are widely spread over the sieve bottom, they should first be concentrated at one edge of the sieve by gently swirling it in about an inch of water. Then tilt the sieve over a wide-mouthed jar and gently "back-pour" the water through the mesh, flushing the materials into the jar. An extra, empty bottle or jar should always be carried into the field for this purpose. No sand, stones or other hard particles should be decanted from the bucket into the sieve or from there into the jar. After being properly labelled, the sample is brought to the laboratory where the organisms are sorted alive from the light trash (leaves, roots, plant stems, algae, etc.) and wet weight or volume is determined.

Field sheets for recording the conditions under which each sample is taken are usually made up in advance of the field work. These may have the following column headings: sample number, date, time of day, air and water temperatures, width of stream, depth at point of sampling, type of bottom, speed of current (in feet per second; determined with a float or current meter in water over area sampled), name of the person doing the sampling, wet weight in grams or volume in cc of the organisms taken, and remarks. The latter provides space for comments on miscellaneous observations such as color of water, turbidity, abundance of aquatic plants, evidence of pollution, or other factors. The field data sheets should be completed for each sample immediately after it has been collected.

Samples should never be taken from disturbed riffle bottom areas. Do not walk over the area to be sampled or take samples from places where cattle have trampled the bottom, or where wading has modified its normal character. Stream faunas are fragile organisms and the slightest disturbance can cause great changes. This is the reason severe floods at times will almost completely denude streams of their macrofaunas.

Sorting the animals from the light trash and taking their weight or volume is the next step. Hand picking of each individual organism from trash is a time-consuming job but essential for quantitative work. Various methods have been tried to find a simple and quick method for separating organisms from the trash taken in bottom samples, but to date none has been successful.

In the most efficient method now available, water is whirled in a jar and, while in motion, decanted in small quantities into a white enamel pan. Add clean water to a depth of about half an inch and start picking the organisms out with a pair of forceps or small lifting screen (fig. 6). They should be placed in another tray of clean water free of all debris, from which they will be weighed. Only a small portion of the sample should be picked at a time because large amounts of debris make them difficult to see. Clean water may be added to the sample jar to facilitate the separation of animals from inert materials.

Weights or volumes must be determined, as counts alone of the organisms in bottom samples do not give an adequate picture of abundance of aquatics. Many samples will contain large numbers of small animals and their total weight will be small. Samples that contain crayfish or the larger caddis-worms, stonefly or mayfly nymphs, add materially to the weight and offer much more, in the way of fish food at least, than the smaller but more numerous forms. Fishes or salamanders taken in a sample should be discarded; these are not part of the bottom fauna.

In samples containing snails, clams or case-bearing caddis-worms, either the cases or shells should be removed before weighing or a factor applied to reduce the weight to that of the animal bodies alone. In practice we have found it best to remove caddis cases and to weigh the molluscs separately, recording these figures as parts of the sample total.

A sensitive torsion balance is used to determine the total wet weight in grams of the organisms. Organisms are recovered by pouring the collection from the sorting tray over a fine-meshed screen such as an ordinary tea strainer. The screen and its retained organisms are then held on blotting paper for a standard time of one minute to remove excess moisture. The organisms are then transferred to a weighing pan (or better, to a tared glass container) by inverting and sharply tapping both the strainer and the receiver against a table top. To get the few animals that adhere to the strainer, flick the strainer sharply against the wetted bottom of a white enamel tray; on it, they can easily be seen and picked out. Prompt weighing minimizes water losses by evaporation and standardizes the operation. The weighed samples are preserved in 70 per cent ethyl alcohol, labelled and held for analysis of the genera and species collected. The wet weight in grams is entered on duplicate field sheets opposite the proper sample number. We usually make a carbon copy of all data sheets and keep them in separate places so that if one is lost or becomes misplaced, another is available, and the results of the work are not lost.

To determine volume of a sample much the same procedure is followed. Drain the organisms on blotting paper for one minute and then drop them in a graduated, centrifuge tube filled to a given level with water. Record the volume in cubic centimeters by noting the rise in level. Wet weights are preferable to volume because the figures obtained are readily converted to grams per square meter, pounds per acre or any other unit area.

There is a great advantage in picking out the organisms when alive rather than preserving them and picking them over later. Alive, the insects move about and are easy to see. When dead, many resemble bits of trash and the sorting job becomes slower and more tedious and inaccurate.

Lake bottom sampling methods. Sampling static water, bottom faunas requires use of the Ekman dredge (fig. 12) or other suitable apparatus. The Ekman dredge consists of a six-inch, square brass box fitted with spring-operated closing jaws. The jaws are cocked open at the surface by chains attached to a release mechanism. It is then lowered by a rope to the bottom. Here, after

SUGGESTED DATA SHEET

for recording the numbers and stages of each order of organisms collected in quantitative stream or lake bottom samples

Stream Bottom Sample

Name of Stream _____ Station No. _____ Date _____
Sample No. _____ Flow/sec _____
Width: _____ Temp: Air: _____ Water: _____
Depth: _____ Time: _____
Type of Bottom: _____ Recorded by: _____
Wet wt. of food: _____ gms.

Order	Total No.	Stage*	Per cent	Remarks
Trichoptera				
Ephemeroptera				
Diptera				
Plecoptera				
Coleoptera				
Neuroptera				
Crustacea				
Misc.				
Totals				

Lake Bottom Sample

Sample No. _____ Station No. _____ Date _____
Name of Lake _____ R. _____ T. _____ S.
County _____ Type of Bottom _____
Temp: Air: _____ Water: _____ Depth _____
Time: _____ Vegetation _____
Wet wt. of food: _____ gms. Distance from shore: _____
Recorded by: _____

Order	Total No.	Stage*	Per cent	Remarks
Molluscs				
Ephemeroptera				
Diptera				
Oligochætes				
Coleoptera				
Neuroptera				
Crustacea				
Misc.				
Totals				

* Record as *L* for larvæ; *N* for nymphs; *P* for pupæ; *A* for adults.

the bottom is felt, it is raised a foot or so upwards and held a moment to permit the line to straighten out for a direct vertical fall, after which it is allowed to drop quickly into the bottom. A metal messenger is then sent down the rope which trips the spring release, closing the jaws. Next, it is pulled to the surface but *not* raised above the surface of the water. This is important for securing accurate samples without spill or leakage of organisms when transferring the sample from the dredge into a bucket. Transfer is accomplished by forcing into the water surface a bucket tilted sidewise toward the dredge and at the same time lifting the Ekman into it. Water will flow into the bucket as the dredge is lifted so that any over-spill from the top goes into the bucket with the water. Since the bottom silt taken with the organisms will pass easily through a 28-mesh sieve, the next step is to stir the sample, mixing the silt and water, and pour it through the sieve held in the water surface. This prevents injury to the soft-bodied midge larvae and pupae and other forms taken. Water is added to the bucket as needed to dilute the bottom silt in order to pass it through the sieve. Once the sample is concentrated in the sieve it is transferred to a jar in exactly the same manner used with Surber samples, namely, by "back-pouring" through the sieve and washing the materials into a wide-mouth jar. Subsequent treatment is the same as that described for Surber samples.

In taking Surber or Ekman samples, it is most important to make sure that none of the organisms is lost at any step in the processes of concentrating, decanting, sorting, weighing, or bottling. If spill occurs at any step, the sample should be discarded. Quantitative samples inaccurately taken are worse than none at all.

DISTRIBUTION AND ABUNDANCE OF STREAM-DWELLING ORGANISMS

Studies by the junior author have clearly indicated that the riffles are the "larders" of streams while pools are usually much poorer in food. This is the reason most sampling to determine the abundance of aquatic organisms is done in riffles. Pools usually provide much more shelter for fishes than riffles but they are often scoured by flood waters, which greatly reduces their productivity. In contrast, riffles usually provide a far greater variety of habitat niches for aquatics, swifter currents for insects that strain their food from the water, and shallow, well-lighted areas that permit photosynthesis to operate more effectively in producing plankton organisms upon which many aquatic animals feed. Back eddies or settling basins in pools may temporarily harbor rich assortments of organisms but such conditions are highly unstable. When the substrates of riffles are composed of large or medium-sized rubble, they present probably the most stable type of stream environment. Even so, flood and drought, ice and snow, coupled with high summer temperatures in many waters, give stream inhabitants an extremely precarious existence.

It should be made clear that stream faunas are enormously variable in both their distribution and seasonal or annual abundance. In the main courses of larger streams, the factors that cause such wide variation in numbers of aquatics are the ones that give streams their reputation for high instability: namely, floods, droughts, ice, and snow, along with wide daily and seasonal changes in water temperatures. Samples taken in Waddell Creek near Santa Cruz, California, one fall, gave an estimated standing crop of over 200 pounds (wet weight) per acre. Only two months later following a severe flood, a similar series of samples taken in the same riffles gave an average of only 70 pounds per acre. Screening the water during the flood revealed many pieces of mayfly and stonefly nymphs, caddis larvae and pupae and other forms that had been ground to bits by the stones being washed downstream. Later in May, the same riffles gave a standing crop of over 470 pounds per acre. One major reason for this great increase in the spring was that the blackfly eggs had hatched in the meantime and hordes of their larvae covered the rocks in swift water. The eggs of other forms were hatching too and the net result was a vast increase in their quantitative abundance over a fairly short period of time.

Annual variations in the abundance of aquatics are equally high. Over a five year period Convict Creek in eastern California varied from a low of 68 pounds per acre to a high of 197 pounds per acre.* These figures represent summer-time averages. Such wide fluctuations in the supply of aquatics clearly indicate that the number of pounds of fish that might be grown each year from such a stream would certainly show similar variability.

One frequently hears it stated that fish should not be planted in the fall because of lack of food in streams in winter. This is erroneous. In streams subject to severe winter conditions aquatic organisms are more abundant in the substrate in winter than they are in summer. Four samples taken in the Merced River in August produced an average of 85 pounds, while samples taken in February when heavy snow and ice covered the stream gave a figure of 103 pounds per acre. The greater winter abundance of aquatics is easily explained by the fact that the entire spectrum of insects dwelling in any given stream has to be present in the winter. In the warmer months many species will have emerged as adults to mate and to lay eggs and may not be present as nymphs, larvae or pupae when the sampling is done. Only a few kinds are fitted for emergence in cold weather, so the greatest aggregations are present in winter.

In a single, square-foot sample taken in the Klamath River near Hornbrook, California, over 4,000 living organisms were collected, giving a standing crop of over 5,000 pounds per acre. The reason this enormous productivity of almost one organism for every two square millimeters of bottom area is this: Copco Lake above Hornbrook produces in the summer tremendous quantities of microscopic plankton organisms which are constantly flowing from the dam into the river below. As a result, aquatics like the net-spinning caddis larvae, aquatic moth larvae, and other forms that feed on these materials, find ideal conditions for existence and develop enormous populations accordingly.

Contrasting with the highly unstable lower main reaches of streams are small, quiet, spring-fed, headwater tributaries. The latter probably represent the

---

* From John Maciolek and P. R. Needham, 1951, *Ecological Effects of Winter Conditions on Trout and Trout Foods in Convict Creek, California,* Trans. Amer. Fish., Soc., Vol. 81, pp. 202-217.

most stable type of stream habitat in the world. They are characterized by even flows and temperatures the year round and as a result are extremely rich in food, some of them showing truly tremendous numbers of aquatic animals per acre of bottom area. Hot Creek near Bishop, California, which is one of America's richest spring-fed trout waters, produced an estimated standing crop of almost 10,000 pounds per acre, largely made up of the fresh-water shrimp *Gammarus* and the young of various caddisfiies. The very special aquatic conditions existing in these waters are responsible for such tremendous biological productivity; they are cited to illustrate further the extreme variability encountered. The vast seasonal and annual variations in productivity found from stream to stream would appear to make long-term, established stocking policies for hatchery-reared fishes in streams a highly questionable affair, especially where the same numbers and sizes of fishes are recommended on a continuing basis from year to year.

Because of the extreme variability in the distribution and abundance of stream-dwelling organisms, even in a single riffle, Dr. R. L. Usinger and the junior author* found that 194 samples would be required to give significant figures on wet weights at the 95 per cent level of confidence. Total numbers would require 73 samples. However, it was also found that only one or two samples would be required to obtain at least one representative of the commoner forms of stoneflies, mayflies, caddisflies and true flies present. These findings were based upon 100 samples that were taken from a single riffle. Results obtained with the Surber sampler, while inadequate statistically as to weights or numbers when only a few samples are taken, nevertheless are quite valuable in indicating general levels of abundance. Most such quantitative studies have been made using the Surber sampler, and its use should be continued until a better sampling device is invented, since the results will then be comparable to much other published work. Needless to say, the data presented is the best we have and may prove to be the best that can be obtained because of the vast range of variation encountered in studies of aquatics on a quantitative basis.

ANALYZING WATER FOR OXYGEN, CARBON DIOXIDE, AND ALKALINITIES†

Collection of samples. Bottles for dissolved oxygen samples should be of 250-300 cc capacity and have ground-glass stoppers. A standard short form (33 × 200 mm) model 100 cc Nessler tube should be used for carbon dioxide determination. Water samples for the determination of alkalinity may be measured with a 100 cc pipette or volumetric flask into a 250 cc Erlenmeyer flask for titration.

In streams, surface samples for dissolved oxygen may be collected by tilt-

---

* Needham, P. R. and R. L. Usinger, 1956, *Variability in the Macrofauna of a Single Riffle in Prosser Creek, California, as Indicated by the Surber Sampler.* Hilgardia, Vol. 24, No. 14, Univ. of Calif., pp. 383-409.

† Modified from H. S. Davis, 1938, *Instructions for Conducting Stream and Lake Surveys,* Fishery Circular No. 26, U.S. Dept. Commerce, Bur. of Fisheries, U.S. Govt. Printing Office, Washington, D.C., 55 pp.

ing the bottle to a horizontal position and allowing the water to flow into it very slowly and gently. Care should be exercised to prevent undue aeration and the bottle should be filled so that water is displaced when the stopper is inserted and no air bubbles are left under it. Air bubbles may be avoided to a great extent by first rinsing out the bottle to moisten the interior surface.

The carbon dioxide sample may be collected in a similar manner in a low form Nessler tube. The water level should be accurately adjusted to the mark by gently pouring off the surplus without undue agitation. This method of collecting samples is suitable for waters which are not heavily saturated or badly polluted, and where a high degree of accuracy is not required.

Samples from various depths may best be collected in a standard Kemmerer sampler (fig. 13). A rubber tube connected to the drain valve at the bottom of the sampler may be used to empty the sample into suitable containers. The tube is inserted in the bottom of the vessel and the water allowed to flow until the contents of the vessel have been changed twice.

Since no exchange of gases ordinarily affects the alkalinity of the water, it is not necessary to exercise as great care in the collection of the samples such as is required for dissolved gases. All procedures should be carried out according to "Standard Methods of Water Analysis," eleventh edition.

Fig. 13. Kemmerer water sampling bottle in open position, showing messenger that closes it.

Reagents

1. Manganous sulphate: 480 grams $MnSO_4.2H_2O$ dissolved in distilled water; filtered and made up to 1 liter.

2. Alkaline-iodide: 500 g. NaOH or 700 g. KOH (these may be used interchangeably) and 135 g. NaI or 150 g. KI in distilled water made up to 1 liter.

3. Concentrated sulphuric acid: sp. gr. 1.83-1.84.

4. Sodium thiosulphate: Use 6.205 g. $Na_2S_2O_3$ in distilled water; make up to 1 liter. This makes an N/40 solution. Add 5 cc chloroform to preserve it. Use new solution every 3-4 weeks. Standardize occasionally against N/40 potassium dichromate solution as directed below.

5. Starch solution: Dissolve 5 g. of potato starch in small amount of distilled water and make up to 1 liter. Starch solutions, even if preservatives such as chloroform or zinc chloride are added, deteriorate quite rapidly, especially in warm weather. A satisfactory method is to use sterilized solutions in small bottles which are opened as required.

6. Phenolphthalein indicator: Dissolve 2 g. in 400 cc of 50 per cent alcohol. Neutralize with N/50 sodium hydroxide. Use boiled distilled water to dilute the alcohol.

7. Sodium hydroxide: N/44 solution preferably made up in some properly equipped chemical laboratory. Since N/44 hydroxide, or stronger solutions, deteriorate once the bottle is opened, this should be restandardized, or a fresh one obtained at frequent intervals. It is helpful to have this reagent supplied in small, tightly corked containers.

8. Methyl orange indicator. Dissolve 0.2 g. in 400 cc of distilled water.

9. Sulphuric acid, 0.2 N or N/50. Make up according to standard specifications.

Untrained persons should not be permitted to make oxygen analyses. Care should be taken in handling the concentrated sulphuric acid. If spilled on clothes it will quickly eat holes in them and it is highly irritating to the skin. It should be carried inside a lead container to avoid damaging leakage. In drawing it up in a pipette extreme care should be exercised to keep the tip of the pipette well below the surface in the acid. This will avoid entraining air bubbles and drawing it into the mouth where it will cause serious burns.

Method of Standardizing Sodium Thiosulphate Solution Used in Oxygen Determinations

1. Dissolve 2.5 grams of KI (potassium iodide) in 50 to 75 cc of distilled water in an Erlenmeyer flask.

2. Add 0.5 cc of concentrated sulphuric acid.

3. Add 20 cc of .025 N potassium dichromate solution from a burette.

4. Cover the mixture with an opaque cloth, or place it in a dark place, for 5 minutes.

5. Titrate with the sodium thiosulphate to be standardized, using a few drops of starch solution as the indicator.

If the thiosulphate is exactly N/40, the desired normality, then exactly 20 cc will be required in the titration and the titration value (cc) is equal to the oxygen present in parts per million (ppm). If more or less than 20 cc is used, a correction factor may be easily obtained as follows:

a. If more than 20 cc is used, the solution is weaker than N/40 (.025N) and the titration value will be too high. The correction factor therefore should be less than 1. For instance, if 22 cc is used, divide 22 by 20, which gives 0.91. This is the correction factor. If the titration value for oxygen comes out 12 cc, multiply this figure by 0.91 to get the corrected value of 10.9 ppm.

b. If 18 cc is used in the standardization, the sodium thiosulphate is more concentrated than N/40 and the titration value will be too low. Again dividing: 20/18 = 1.11 (factor). If the oxygen determination comes out 8 cc, multiply this by 1.11 to get a corrected value of 8.88 ppm of oxygen.

Dissolved oxygen:

1. To the sample collected in the 250-300 cc ground-glass stoppered bottle, add 1 cc of manganous sulfate and 3 cc of alkaline potassium iodide beneath the surface of the liquid.

2. Replace the stopper, at the same time preventing the entrapping of air bubbles. Shake vigorously for 15 seconds and allow the precipitate to settle until it is contained in the bottom half of the bottle. If this does not occur rapidly the shaking should be repeated.

3. When the precipitate has settled add 2 cc of concentrated sulphuric acid above the water level, holding tip of pipette against the neck of the bottle. Replace the stopper, taking precautions against trapping air bubbles, and shake the bottle vigorously to mix the contents. Up to this point the procedure must be carried out in the field. The final titration should be made as soon as possible thereafter, but a lapse of a few hours is permissible.

4. Transfer 200 cc of the treated sample, measured with a 200 ml volumetric flask, into a 500 cc Erlenmeyer flask. This is conveniently accomplished by carrying into the field a 200 cc flask cut off a few millimeters above the mark and inserted into a pierced stopper which fits the neck of the larger flask. By inserting the mouth of the smaller flask into the larger flask and reversing the two with a slight rotary motion the sample will transfer rapidly.

5. Titrate the sample with 0.025 N(N/40) sodium thiosulfate. One or two cc of starch solution should be added only when the color has become a faint yellow after the addition of thiosulfate. Titrate over a white background until the blue color disappears.

Dissolved oxygen is reported in parts per million by weight. If a 200 cc sample is used the dissolved oxygen in parts per million is equal to the number of cc of 0.025 N thiosulfate required. No corrections are necessary for the reagents added except in work of unusual precision.

Carbon dioxide. Collect 100 cc of the sample in a low form Nessler tube (200 × 32 mm) according to one of the two methods outlined. Add 10 drops of phenolphthalein and titrate rapidly with N/44 sodium hydroxide until a faint but permanent pink is produced (3 minutes). The sample may be mixed by swinging the container or rotating it with a circular motion of the wrist. Any agitation of the surface of the liquid tends to change the gaseous content and should therefore be avoided. The free carbon dioxide is equal to 10 times the number of cubic centimeters of N/44 sodium hydroxide used.

Alkalinity. The alkalinity of natural waters represents its content of carbonates, bicarbonates, hydroxides, and occasionally borates, silicates, and phosphates. Alkalinity is determined by titration with a standard solution of strong acid to certain datum points or hydrogen-ion concentrations. Indicators are selected which show definite color changes at these points.

Since dilute bicarbonate solutions have a hydrogen-ion concentration of about pH 8.0 and dilute carbonic acid solutions a hydrogen-ion concentration of about pH 4.0, these are chosen as datum points, and indicators should be selected which show definite color changes at about these two points. The amount of standard acid required to bring the water to the first point measures the hydroxides plus one-half the normal carbonates (phenolphthalein); the amount needed to bring it to the second point corresponds to the total alkalinity (methyl orange).

The sample may be collected by any method since there is no great danger of changing the alkalinity. Both end points may be determined on the same sample by using phenolphthalein first and continuing with methyl orange after the first end point has been determined.

1. Phenolphthalein alkalinity. Add 4 drops of phenolphthalein indicator to 50 or 100 ml of sample in a white porcelain casserole, or an Erlenmeyer flask over a white surface. If the solution becomes colored, hydroxide or normal carbonate is present. Add 0.02 N sulfuric acid from a burette until the coloration disappears.

The phenolphthalein alkalinity in parts per million of calcium carbonate is equal to the number of milliliters of 0.02 N acid multiplied by 20 if 50 ml of sample was used, or by 10 if 100 ml was used.

2. Methyl orange alkalinity. Add 2 drops of methyl orange indicator to 50 or 100 ml of the sample, or to the solution to which phenolphthalein has been added, in a white porcelain casserole, or an Erlenmeyer flask, over a white surface. If the solution becomes yellow, hydroxide, normal carbonate, or bicarbonate is present. Add 0.02 N sulphuric acid until the faintest pink coloration appears, that is, until the color of the solution is no longer pure yellow.

The methyl orange alkalinity in parts per million of calcium carbonate is equal to the total number of milliliters 0.02 N sulfuric acid used multiplied by 20 if 50 ml of sample was used, or by 10 if 100 ml was used.

Hydrogen-ion concentration. The *p*H, or hydrogen-ion concentration, is a measure of the degree of acidity or alkalinity of a solution. It is most accurately measured by means of a potentiometer. Many types of simple *p*H kits are available today for measuring this factor under field conditions.

A *p*H value of 1 to 7 is acid, 7 represents the neutral point, while values from 7 to 14 are alkaline. To measure *p*H, a few drops of indicator solution are added to a small sample of the water and the resulting color is compared to standard color charts or colored glass discs provided for various *p*H ranges. The following indicators with their *p*H ranges are listed as follows:

Indicator	*p*H range
Brom cresol purple	5.2-6.8
Bromthymol blue	6.0-7.6
Phenol red	6.8-8.8
Cresol red	7.2-8.8
Thymol blue	8.0-9.6

Care should be taken not to contaminate the sample by inverting the tube against the finger tip to mix indicator and sample. A rotary movement with the wrist will suffice.

METHODS OF COLLECTING FISHES

The age-old method is with rod and line. Sometimes it is the only way certain fishes can be caught, but most of the time it is too slow and is highly selective, catching only those forms that will bite readily on natural or artificial lures. Fishes have been taken by many methods such as dynamite, quicklime, traps, poisons, shooting, guddling, spears and spear-guns, and electric shocking. Most of these methods are now illegal. A person wishing to collect fishes for teaching or research purposes should secure a scientific collecting permit first from the appropriate state or federal agency controlling the waters of the area in which collections are to be made.

The best collecting methods are those of the commercial fisherman, who uses gill nets, seines, and traps. His minnow pails and live boxes are also the best means of keeping living material. In addition, rotenone (also called cubé or derris root) powder may be used to treat chemically small water areas. This causes asphyxiation of the fishes, and they may be easily lifted out of the water with a dip net when they come to the surface for air.

The safest, cheapest, and easiest way to obtain fishes for study is to use small minnow seines, ¼ inch mesh, 10 feet by 4 feet. With stout, wooden brails tied on each end, two men seining (one working each brail) can quickly obtain good representative collections. For seining over wider areas of water and securing larger fishes in greater numbers, a beach seine is useful. These can be secured in any desired size, but a net with ¼ inch webbing, 100 feet long by 6 feet deep, is ample for most purposes.

Field Sheet For Fish Collection Notes

State or Country _____   Field No. _____

County _____   Date _____

Locality _____   Collector(s) _____

Stream or Lake _____   No. of Specimens taken _____

Stream (name) _____   Lake (name) _____

Width _____   Area (if known) _____

Depth _____   Drainage System _____

Current Speed _____   Water Color _____

Temp. Air _____ Water _____   Shore _____

Time[1] _____ A.M. _____ P.M.   Cover _____

Shelter _____

Vegetation _____

Bottom _____

Turbidity (if any) _____

Depth of Capture _____

Method of Capture _____

Preserved in _____

[1] Always record time of taking air and water temperatures. Air temperatures should be taken in the shade of some object, never in direct sunlight.

PRESERVATION OF FISHES

Fishes are best preserved in 10 per cent formalin (ten parts of water to one part of formalin). Household borax (one teaspoon per quart of the preserving solution) will retard shrinkage of the specimens and aid in preserving them.

Specimens larger than three inches should be slit on the right side to permit penetration of the preservative around the internal organs. The slit should be at least one-third as long as the body cavity. The right side is recommended for this since the left side is used for various measurements and is usually shown in photographs.

Fixation by formalin may take from a few hours with small specimens to a week or more with large forms. After fixation the specimens should be washed thoroughly in running water or by several changes of water for a period of at least 24 hours, after which they should be placed in 40 per cent isopropyl alcohol. One change of alcohol is necessary to remove the last traces of formalin; thereafter they may be permanently preserved in the 40 per cent isopropyl alcohol.

Fish fade after preservation. Unfortunately there is at present no known way in which to preserve the colors of fishes. For this reason, it is important to record color notes at the time of collection. Color photographs taken promptly after the fish are removed from the water and before preservation will aid materially in securing adequate color notes.

Besides color notes on the fishes themselves, adequate field notes should be taken on each collection made. A sample form for field notes is given on p. 94. Additional notes can be made on the reverse side. All writing should be done with a waterproof carbon ink that produces a permanent record. Copies of field notes should always accompany any given collection. The label placed inside the container with the specimens should bear the same number or designation as the field notes, including the locality data, date, and collector's name. A numbered tag placed on the outside of the container will make it easy to pick out any single collection without having to find the label inside the container. Any detailed ecological, behavioral, or other notes should accompany the field sheet for each collection.

Where lack of containers forces the use of a single container for several collections, each collection should be wrapped separately in a cheesecloth bag after placing the proper label inside. Such bags are best sewed shut, for they often become loosened in transit and mix together irretrievably fishes from different collections.

If the specimens are too large for wrapping, individual labels can be placed in the throat or gill chamber of each fish and removed later in the laboratory. Unlabelled specimens lacking adequate field data are of little scientific value though they may be useful for other purposes.

A sample of a fish stomach analysis form that has proven useful is presented on p. 96.

FISH STOMACH ANALYSIS RECORD

Data Recorded by _____
Date _____

Species _____ Sex _____ Total length in mm. _____
Date _____ Lake _____
Stream _____ Township _____
Section _____ Section _____
County _____
Range _____

Order	No. Individ.	Vol. in cc	% of Total Vol.	Percent by No.	No. Individ.	Vol. in cc	% of Total Vol.	Percent by No.	No. Individ.	Vol. in cc	% of Total Vol.	Percent by No.
Ephemeroptera												
Trichoptera												
Diptera												
Coleoptera												
Plecoptera												
Homoptera												
Hemiptera												
Arachnida												
Hymenoptera												
Miscellaneous												
Total												

METHODS OF COLLECTING FISH SCALES

Information on the age and rate of growth of fishes is of considerable interest biologically and is important in relation to fish stocking and management. The "rings" or circuli laid down on the surface of scales reflect growth history and can be read much like the rings on the stump of a tree to determine age. Structures such as otoliths (ear bones) and other bones are also used, but the scale method has proven the most useful for determining the ages of most fresh-water gamefishes.

The best place from which to take scales is the midregion of the body immediately above the lateral line. Scales should not be taken from the lateral line. With fish over 12 inches long, it is well to wipe off mucus with a cloth before scraping the scales loose, as large amounts of mucus make it difficult later to clean the scales for mounting. Removal is effected by scraping against the outer free tips of the scales with a dull knife or scalpel.

After removal, scales should be placed in regular scale envelopes, which are available for this purpose. The mucus taken with the scale samples is usually ample to attach them firmly in place. Extreme care should be taken to prevent the mixing of scales from different fish. The knife should be wiped clean after each sample is taken. Complete data as to date, name of water, species of fish, sex, total length, and collector, should be recorded with each sample taken. Where condition factors are also to be determined from the same fish, it is highly desirable that the fish be weighed fresh, as preservation in formaldehyde changes the weights markedly.

Scales collected obviously cannot be studied in the field but should be retained for later study in connection with the write-up of the stream and lake survey findings.

STREAM AND LAKE SURVEYS

Surveys of the waters in any given area offer the principal means of obtaining preliminary basic biological, physical, and chemical knowledge of our waters. They also provide the fishery manager with data essential to the development of adequate fish stocking and management policies. For the teacher, they provide knowledge of individual waters and their character and make known places to take classes for studies of specific types of aquatic habitats. We have found wonderfully rich waters where field trips could be held within short distances of schools. The field forms used in reporting stream and lake survey data are given on pages 98-100.

REFERENCES

Carlander, Kenneth D.
  1950   *Handbook of freshwater fishery biology*. Wm. C. Brown Co., Dubuque, Iowa, 286 pp.
Lagler, Karl F.
  1959   *Freshwater fishery biology*. Wm. C. Brown Co., Dubuque, Iowa, 421 pp.

STREAM SURVEY FORM

1. State   River system   Name of stream
      Forest or park   Map   Number
      County   Tributary to
      Stream section:
        From:
        To:
        Length of section:
     Notes.—Sketches (show trails, roads, tributaries, stations, barriers, springs, etc.):

2. Name of stream _____   Date _____

Region	Upper	Middle	Lower
Station			
Altitude			
Average width and depth			
Volume			
Velocity			
Color and turbidity			
Alkalinity			
pH			
Air temperature			
Water temperature			
Hour and sky			
Pools:			
Size, type, frequency	S T F	S T F	S T F
Caused by			
Shelter			
Bottom type:			
Pools			
Riffles			
Shade			
Aquatic vegetation			

Fish food (per sq. ft.):
   Caddisflies   Beetles
   Mayflies   Other insects
   Diptera   Crustacea
   Stoneflies   Miscellaneous
Vol. in cc per sq. ft.
Character of watershed: Canyons, mountainous, hilly, rolling, flat, swampy, wooded, open
   cultivated, uncultivated.
Character of subsoil, bedrock and dip of strata.
Condition of stream: Low water, normal water, high water
Fluctuation in volume:
Gradient:
Source:
Barriers (type, location, height):
Diversions (type, location):
Springs (location, volume, temperature):
Tributaries (number and size):
Fish (kinds, av. size, abundance, stations):
Enemies:
Degree fished (heavy, medium, light):
Spawning areas:
Fry, fingerlings seen (kinds, abundance, stations):
Accessibility of stream (by car or . . . miles by trail):
Previous stocking:

Pollution (source, type):
Rearing pool sites:
Fish recommended:
   Species:
   Reasons:
Remarks:
Improvements:

_____
            Investigator

Average width: Pool grade: Food Grade:

Stocking Program

Section to be stocked:
Species:
Size:
Number:
Frequency:

_____
            Authority

LAKE AND POND SURVEY FORM

1. State   River system   Name of lake
   Forest or park   Map
   County   Tributary to   Number
   Notes.—Sketches (show soundings, stations, weed beds, tributary streams, outlet, roads, and trails on outline map):

```
          N
          ↑
         Map
```

2. Name of lake: Date:
   Altitude: **Area:**
   Natural or artificial:
   Height of dam: Fishway:
   Character of shore line: Rocky, boggy, sandy, muddy, meadow, wooded.
   Character of watershed: Mountainous, hilly, rolling, flat, swampy, wooded, open, culti-
      vated, uncultivated.
   Principle tributary streams (number and size):
   Fluctuations in water level (causes and feet variation):
   Approximate depth 100' from shore _____, 200' from shore _____
      Maximum _____
   Shoal areas 20' or less _____ % of lake:
      Bottom—Mud, silt, sand, clay, peat, marl, detritus, hardpan, gravel, bedrock.
   Deep areas: Bottom—Mud, silt, sand, clay, peat, marl, detritus, hardpan, gravel, bedrock.
   Temperatures: Inlet _____, Outlet _____, Surface _____
      Air _____, Hour _____, Weather _____
   Color: Turbidity:

Higher plants (show location on map):   Abundance
   Emergent:
   Submerged:
   Algae (kinds):   Abundance
   Vertebrates:
     Kinds of fish: Abundance
     Other vertebrates:
   Invertebrates:
     Shore (approx. no. per sq. ft.): Stoneflies _____, Mayflies _____,
       Caddisflies _____, Odonata _____, Diptera _____ , Snails _____,
       Amphipods _____, Miscellaneous _____.

Open water stations	1	2	3	4	5
Depth in feet:					
Plankton					
Length of haul					
Quantity in cc					
Bottom (per ¼ sq. ft.):					
Midges					
Annelids					
Snails					
Clams					
Type of sample bottom					

Amphipods
Miscellaneous
Volume in cc

Spawning areas:
Young fish seen:
Accessible by car or _____ miles by trail.
Boats available:
Polution: Source:
        Type:
Degree fished (heavy, medium, light):
Rearing pool sites:
Fish recommended:
   Species:
   Reasons:
Remarks:
Improvements:

                                     _____
                                        Investigator

Stocking Program
   Species:
   Size:
   Number:
   Frequency:

                                       _____
                                        Authority

STUDIES OF AQUATIC ENVIRONMENTS

Aquatic environments may be roughly divided into two general types, lentic and lotic.* Organisms dwelling in the former—quieter waters of lakes or streams—show no special adaptations for withstanding the wash of waves or currents, whereas organisms living in moving waters show modifications in form and habits to adapt to their environment. It is not always possible to draw a clear line between the two situations, and some organisms transgress from one to the other. Both types offer rich sources of living material for study.

Aquatic habitats also may be divided roughly into the following three categories on the basis of their water supplies: (1) habitats with permanent water; (2) those that dry up occasionally; and (3) those that have water at rare intervals. All three types intergrade with years of scarcity or abundance of rainfall.

> Perhaps . . . for bodies of still water the words pond, pool and puddle convey a sense of their relative permanence. . . . The population of the pond is, like that of the lake, to a large extent perennially active. . . . That of the pool is composed of those forms that are adjusted to drought: forms that can forefend themselves against the withdrawal of the water by migration, by encystment, by desiccation, by burrowing, or by sending roots down into the moisture of the bed. . . . The puddles have a scanty population of forms that multiply rapidly and have a brief cycle. The synthetic forms among them are mainly small flagellates and protococcoid green algae. The herbivores are such short-lived crustaceans as *Chirocephalus* . . . and *Apus*, which have long-keeping, drought-resisting eggs; such rotifers as *Philodina*, remarkable for its capacity for resumption of activity after desiccation; such insects as mosquitoes. The carnivores are such adult water-bugs and beetles as may chance to fly into them.†

Populations of lotic waters are also perennially active below levels of minimum flow in permanently wetted areas. Whether the water flows in many directions as on lake shores or steadily in one direction, as in streams, makes little difference; the organisms inhabiting each will be much of the same kinds. The plants will be mainly such tough, current-resisting forms as the alga *Cladophora* and the diatoms that form exceedingly slippery coats of slime on the bottom stones and gravels.

From the standpoint of their food supply, two main groups are recognized: (1) those that feed on plankton by means of tiny nets or other apparatus for straining minute organisms from the currents of water; and, (2) ordinary feeders that forage freely over the bottom eating smaller animals or plants.

The animals of lotic waters are mainly small invertebrates. Those most easily seen and collected are principally the larvae and pupae of caddisflies and aquatic true-flies, stonefly and mayfly nymphs, and the larvae of various dobsonflies, alderflies, hellgrammites, beetles, and moths.

---

* *Lentus* = calm, placid; *Lotus* = washed.
† Needham and Lloyd, *Life of Inland Waters.* Comstock Pub. Co., Ithaca, N.Y., 1916, 438 pp.

Suggested studies of fresh-water habitats

A list of suggested studies is given below. About half of these are association studies dealing with the macrofaunas present in various types of aquatic habitats. The remainder deal with specific functions of various structures found in aquatic animals or problems concerned with their management for the benefit of man.

The list is purely suggestive. Many other types of exercise could be included, such as studies of hot springs, vernal pools, rock surfaces in spray-zones, swamps, salt-marshes, and brackish water estuaries. All of them become more meaningful if combined with field trips where the student can view each type of aquatic habitat and aid in the collection of materials for the indoor laboratory to follow.

1. A pond
2. A rapid stream
3. A slow stream
4. A lake shore
5. Lake or pond plankton
6. Stream plankton
7. Flood-plain marsh
8. A bog pond
9. Rock ledges in a stream bed
10. Cold, spring-fed brook
11. Blanket algae association
12. Littoral areas of reservoirs having highly fluctuating water levels
13. An intermittent stream when it is flowing
14. A polluted stream
15. Hibernating devices of aquatic animals
16. Respiration gills of aquatic insects
17. Adaptations for flotation and swimming displayed by aquatic animals
18. Food of fishes
19. Fecundity of fishes
20. The spawning of fishes
21. Stream and lake surveys
22. Fish parasites
23. Age and growth of fishes from scale studies
24. The management of farm ponds
25. Dynamics of fish populations

Outline for studies

Previous editions of this *Guide* contained 25 practical exercises for the use of classes. This fifth edition gives instead a single, simplified outline that we have found useful over the years. It is designed to point up significant items for the student and is based upon the plan of a field trip to a selected type of habitat, followed by an indoor laboratory period where the materials collected are identified and notes concerning them are written up.

SUGGESTED OUTLINE FOR STUDIES

I. Materials needed
List all collecting gear such as dip nets, hand screens, seines, specimen bottles, preservatives, etc., that will be required for the field trip. If weather interferes with plans for a field trip and an indoor laboratory is substituted, then the materials required for any selected indoor study would be listed here. Prepared in advance for any given exercise, such lists have proven to be most useful.

II. Work program
  A. Field trip: List the type of habitat in which collecting is to be done, the location of particular area (pond, lake, or stream, etc.), and the time and place of departure and return.
  B. Laboratory study of the materials collected.
  C. Assigned reading.

III. For the record
For many types of aquatic habitats, the following outline will suggest to the student a suitable format for his notes:
  A. Name of the water, including a brief general description of its salient physical and biological features including such items as temperature, transparency, width and depth (or area), speed of current (if any), type of bottom, etc.
  B. A sketch of the area in which collections were made to indicate ecological zonation if such was evident.
  C. An annotated list of the organisms collected, arranged under the following column headings:
    1. Plants: Notes on species, size, color, abundance, growth habits, etc.
      a. Seedplants—emergent, floating, submerged (in lentic waters) or in current and on stream margins (of lotic waters).
      b. Algae—microscopic, those that color either the bottom or the water itself.
      Fringing (sessile) algae
      Slime-coat or encrusting algae
      Free-floating algae
    2. Animals: Notes on size, abundance, feeding habits if known, stages or ages found, special adaptations or activities, presence of parasites, etc.
      a. Invertebrates
        Free-ranging or free-swimming
        Walking on surface, lying in or suspended from the surface film
        Sessile on vegetation (lentic waters)
        Sessile on stones, logs, roots, etc., or on bottom (lotic waters)
        Burrowers that dig
        Burrowers that squeeze through
        Tube dwellers

MEASUREMENT OF WATER QUANTITIES

The second-foot (cubic foot per second, or c.f.s.) is universally used for determining water quantities.

1 second-foot = 7.48 gallons per second or 448.8 gallons per minute.
1 second-foot = approximately 2 acre-feet per day.
1 second-foot = 86,400 cubic feet per day.
1 second-foot = 646,317 gallons per day.
1 acre-foot = 43,560 cubic feet or 325,850 gallons.
1 cubic foot of water = 7.48 gallons and weighs 62.4 pounds.

WATER VELOCITY AND FLOW MEASUREMENTS[1]

The measurement of water velocity and volume of flow under field conditions may be determined as follows:

Velocity

1. Locate two points 100 feet apart or any convenient distance.
2. Record the time it takes a float to drift between the two points.
3. Compute the number of feet traveled per second by dividing the time in seconds into the distance.

Volume (or rate) of flow

$$\text{Formula: } R = \frac{W\,D\,a\,L}{T}$$

Where R = volume of flow in cubic feet per second.
W = average width of stream in feet.
D = average depth in feet.
a = constant factor for bottom type.
  Smooth sand, etc. = 0.9
  Rough rocks, etc. = 0.8
L = length of stream section measured.
T = time in seconds for float to travel the measured distance.

[1] From *Trout and Salmon Culture,* by Earl Leitritz, California Dept. Fish & Game, Fish Bulletin No. 107, 1959, 169 pp.

EQUIVALENTS

Degrees Centigrade = 5/9 × (°Fahrenheit − 32)
Degrees Fahrenheit = 9/5 × °Centigrade + 32

1 inch = 25.4 millimeters or 2.54 centimeters.
1 meter = 39.37 inches.
1 centimeter = 0.3937 inches.
1 acre = 43,560 square feet or 0.404687 hectare.
1 gram = 15.432 grains.
1 kilogram = 2.205 pounds.

Unit	Gallon	Quart	Pint	Pound	Avoirdupois ounce	Fluid ounce
gallon	1.0	4.0	8.0	8.345	133.52	128.0
quart	0.25	1.0	2.0	2.086	33.38	32.0
pint	0.125	0.5	1.0	1.043	16.69	16.0
pound	0.12	0.48	0.96	1.0	16.0	15.35
ounce	0.0075	0.03	0.06	0.0625	1.0	0.96
fluid ounce	0.0078	0.031	0.062	0.062	1.04	1.0
cubic inch	0.0043	0.017	0.035	0.036	0.573	0.554
cubic foot	7.481	29.922	59.848	62.428	998.848	957.48
cubic centimeter	0.0003	0.001	0.002	0.002	0.035	0.034
liter	0.264	1.057	2.1134	2.205	35.28	33.815
gram			0.002	0.0022	0.0353	0.034

Unit	Cubic inch	Cubic foot	Milliliter	Liter	Gram
gallon	231.0	0.1337	3,785.4	3.785	3,785.4
quart	57.749	0.0334	946.36	0.95	946.35
pint	28.875	0.0167	473.18	0.47	473.18
pound	27.67	0.016	453.59	0.454	453.59
ounce	1.73	0.001	28.3	0.03	28.35
fluid ounce	1.8		29.57	0.03	29.41
cubic inch	1.0	0.0006	16.39	0.0164	16.3
cubic foot	1,728.0	1.0	28,322.0	28.316	28,318.58
milliliter	0.061		1.0	0.001	1.0
liter	61.025	0.0353	1,000.0	1.0	1,000.0

GLOSSARY

(See also the definitions of fish structures beginning on page 66)

ANADROMOUS: ascending from the sea for breeding at certain seasons; e.g., salmon, steelhead, shad, etc.

ANIMALCULE: a minute or microscopic animal

ANTEROLATERAL: situated forward and on the side

APICAL: at, near, or pertaining to the tip or apex

ARCUATE: bowed, arched, or curved

AREOLATIONS: small defined areas, like a network

ARTICULATE: unite by means of a joint

ATRIUM: a cavity, entrance, or passage

AURICLE: an angular or ear-like lobe or process

AUTOSPORE: one of the daughter cells formed by the internal division of one of the unicellular algae duplicating in miniature the parent plant, as in many of the Chlorococcales

AXIAL: of or pertaining to an axis

AXILE: belonging to or situated in an axis

BARBELS: fleshy, slender, tactile appendages around the mouths of various fishes; e.g., sturgeon, catfish, etc.

BASIBRANCHIAL: pertaining to the medium series of unpaired bones behind the tongue in the pharynx of various fishes

BIARTICULATE: two-jointed

BIFID: having two branches or processes

BRANCHIOSTEGAL RAYS: rays supporting the branchiostegal membrane below the gills in certain fishes, e.g. trout, bass, etc.

BUCCAL: pertaining to the mouth cavity

BURETTE: an apparatus, usually glass, for delivering measured quantities of a liquid

CAECA, PYLORIC: pouch-like appendages attached on the outside of the intestine immediately behind the stomach in many fishes

CAUDAL: pertaining to a tail

CEPHALOTHORAX: the united head and thorax of spiders and crustaceans

CERCI: two lateral, anal appendages

CHROMATOPHORE: a pigment-bearing structure or plastid, found commonly in plant cells

CILIA: hair-like processes, found on many cells, capable of a lashing movement

CLYPEUS: that portion of the head of an insect to which the labrum is attached

CONJUGATION: the fusion of two sex cells with ultimate union of their nuclei

CORONA: a crown or crown-like process

COXA: basal segment of the leg of an insect or other arthropod, articulated to the body

CRENULATE: with small scallops, evenly rounded and rather deeply curved

CTENOID: pertaining to a bony type of fish scale with tiny spine-like processes (cf. cycloid)

CULTRATEIFORM: shaped like a pruning knife or a crow's beak

CUNEATE: wedge-shaped

CUPULES: sucker-like processes covering the under surface of the tarsi in male dytiscid beetles

CUTICLE: the outer skin or skin layer

CYCLOID: pertaining to a bony type of fish scale without spine-like processes (cf. ctenoid)

DENDROID: resembling a tree in form

DENTATE: having tooth-like projections

DISCOID: disc-shaped

DISTAL: removed from the point of attachment or origin

ELYTRA: the leathery or chitinous forewings of beetles

EMARGINATE: having a notched margin

ENDOPODITE: inner branch of a typical crustacean appendage

EPIPHYTE: a plant that grows on other plants but is not parasitic on them, e.g., an orchid

ERLENMEYER FLASK: a cone-shaped, flat-bottomed flask

FEMUR: usually the stoutest segment of the leg in arthropods, articulated to the body by the trochanter and coxa and bearing the tibial segment distally

FENESTRATED: having transparent or window-like naked spots, as in the wings of some butterflies and moths

FLAGELLA: whip-like or tail-like processes

FOLIACEOUS: leaf-like

FORMALIN: a preservative solution containing formaldehyde

FURCA: forked anal appendage in some insects, used for leaping

FUSIFORM: spindle-shaped, tapering gradually at both ends

GALEA: the outer lobe of the maxilla in the mouth parts of insects

GLABROUS: without hair or projections, smooth

GLOBOSE: round, spherical

GULAR: pertaining to the throat region

HEPATIC: pertaining to the liver

HETEROCYSTS: large, transparent cells at intervals along the filament in blue-green algae

INTERCALATION: insertion or addition, as a vein in an insect wing

ISTHMUS: fleshy area extending forward on the throat between the gills in fishes

KEEL: an elevated ridge

LABIUM: the lower lip in insects

LABRO-CLYPEUS: the fused upper lip and clypeus in insects

LABRUM: the upper lip in insects

LACINIA: the inner lobe of the first maxilla in insects

LAMELLOSE: made up of or resembling leaves, blades, or plates

LENTIC: calm or placid, as ponds or lakes

LORICA: a hard, protective case or shell (zool.); the cell wall, or two valves, of a diatom (bot.)

LOTIC: flowing, as in river and stream water

LUNATE: crescent-shaped

MANDIBLE: a jaw, either upper or lower

MAXILLAE: the second pair of jaws in insects; bones in the upper jaw in fishes .

MENTUM: a plate usually attached or fused with the submentum on the ventral surface of the labium in some insects

METASTERNUM: the underside or breast of the third thoracic segment in insects

MULTIARTICULATE: having many joints or segments

NESSLER TUBE: a narrow, cylindrical glass jar

OCELLUS: the simple eye in insects

OCCIPITAL: pertaining to the posterior region of the head

ONISCIFORM: shaped like a wood-louse, *Oniscus*

OPERCULUM: (1) the bony covering of the gills in fishes; (2) a covering structure in many insects; (3) plate-like disc on foot of snails (operculate) that seals opening of shell when foot is withdrawn

PALAEOFORM: prehistoric, ancient, or primitive in form

PALATINES: paired bones in the roof of the mouth in fishes, often bearing teeth

PALMATE: like the palm of the hand, with finger-like processes

PALP, PALPUS: one of the mouth parts in insects, a "feeler" with sensory functions

PALPIGER: that portion of the labium bearing the palpi in insects

PAPILLOSE: covered with or bearing small, fleshy protuberances

PARIETAL: pertaining to paired bones in the dorsal surface of the skull of fishes

PATELLIFORM: shaped like a plate, as the modified joint of the anterior tarsi in diving beetles

PECTINATE: comb-shaped or like the teeth of a comb

PECTORAL: the shoulder region just behind gill covers in fishes

PEDUNCULATE: on a stalk or peduncle

PELLICULAR: of or of the nature of a thin skin (pellicle)

PENULTIMATE: next to the last in a series

PINNATE: branched, feather-like

PLASMODIUM: a mass of protoplasm formed by a grouping of amoeba-like cells, such as the body of a slime mold

PLICATE: pleated, folded like a fan

POSTEROLATERAL: situated back and on the side

PREDACIOUS: predatory

PREMAXILLARY: a bone forming the fore part of the upper jaw in fishes

PREOPERCLE: the most anterior of the bony plates which cover the gill apparatus in fishes

PROBOSCIS: an extension of the mouth or nose region, such as the mouth parts of insects or the trunk of an elephant

PROCESS: a part of the mass of an organism that projects outward from the main mass

PRONOTUM: the upper surface of the prothorax of insects

PROTHORAX: the first body segment of insects, to which the anterior pair of legs is joined

PROTRACTILE: capable of being extended

PUPATION: in insects, the stage of undergoing metamorphosis between the larval and adult stages

PUBESCENT: having a hairy covering

PYRENOID: wartlike; a colorless body in the cells of some algae, used for starch deposition

RAMUS: a branch or branch-like structure

RAPHE: a seam or suture such as is found on the valves of diatoms

RAPTORIAL: predatory; adapted for seizing other animals

RENIFORM: kidney-shaped

RETICULUM: network

RETRACTILE: capable of being withdrawn

RIFFLE: a fast-water section of a stream where the shallow water races over stones and gravel

SCLEROTIZED (SCLERITIC): possessing sclerites —hard, chitinous or calcareous plates

SCUTELLUM: the third dorsal sclerite (plate) of the second and third thoracic segments of insects

SERICEOUS: silky

SERRATE: saw-toothed

SESSILE: closely attached; not free to move about

SETAE: bristles or long stiff hairs

SETIFORM: in the form of a seta or stiff hair

SIGMOID: curved in two directions, like the Greek letter sigma or the English S

SILICEOUS: yellow with a slight tint of brown; pertaining to silica

SINUATE: having wavy margins

SPIRACLE: an opening to the trachea or breathing pore in insects; an aperture in some primitive fishes such as sharks and sturgeon

SPORANGIUM: a spore case in which asexual reproductive cells are produced

STERNITE: the ventral part of a ring or segment of an insect

STIGMATA: spiracles of insects

STIPES: the foot-stalk of the mouth parts of insects, bearing the movable parts

STRIATION: a marking of numerous, fine, parallel lines

SUBDICHOTOMOUS: subdivided into pairs

SUBMENTUM: basal sclerite of the labium, by means of which it is attached to the head in insects

SUBSTRATE: base or foundation, such as the soil on which an organism grows

SUTURE: a seam, such as that which joins the body segments of insects

SUPRAMAXILLARY: a bone in the upper jaw lying over the maxillary in fishes

SUBTERMINAL: before the end

TARSUS: foot; the appendage on the legs of insects bearing the claws

THALLUS: the plant body characteristic of certain primitive plants (Thallophytes)

THORAX: the second or middle region of an arthropod's body

TIBIA: the shank of an insect's leg, between the femur and the tarsus

TITRATION: adding a measured volume to a known volume of liquid until a certain effect is observed

TRACHEA: a breathing tube

TRANSECTION: a cut across, at right angles to the body; a transverse section

TRIARTICULATE· three-jointed

TROCHANTER: Second segment in the leg of insects

TROCHANTINE: the basal part of the trochanter when it is two-jointed; the small sclerite connecting the coxa with the sternum

TRUNCATE: cut off squarely at the tip

TUBERCLE: a small protuberance

UMBONES: (1) two movable spines on the side of the prothorax in some beetles; (2) prominences on shells of bivalve molluscs around which growth rings radiate

UNDULATE: wavy or waved

VOMER: a median bone associated with the roof of the mouth, bearing teeth in some fishes

ZONATE: marked with zones; ringed or belted

This fifth edition of *A Guide to the Study of Fresh-Water Biology* is a complete re
book considered almost a classic by biologists, zoologists, and everyone interested in the wild
of ponds, streams, and lakes. Now—with the addition of a section on fishes—it will be of gr
interest to sportsmen as well. With extensive plates of important and commonly found spec
keys for their identification, and discussions of sampling methods, laboratory procedures, a
equipment—this little book is indispensable to anyone interested in fresh-water biology.

HOLDEN-DAY, INC., PUBLISHERS, 500 SANSOME STREET, SAN FRANCIS

Cover design by Jean Swift